58.787
1739
CLC Record

Production Spare Parts

Optimizing the MRO Inventory Asset

Eugene C. Moncrief

Ronald M. Schroder

Michael P. Reynolds

Industrial Press Inc.

Library of Congress Cataloging-in-Publication Data

Moncrief, Eugene C.

Production spare parts : optimizing the MRO inventory asset / Eugene C. Moncrief,
    Ronald M. Schroder, Michael P. Reynolds.

p. cm.

ISBN 0-8311-3228-0

1. Spare parts. 2.Inventory control.--I. Schroder, Ronald M. II. Reynolds, Michael P. III. Title.

TS160.M59 2005

658.7'87--dc22

2005049162

Industrial Press, Inc.
200 Madison Avenue
New York, NY 10016-4078

First Printing, August, 2005

Sponsoring Editor: John Carleo
Interior Text and Cover Design: Janet Romano
Developmental Editor: Robert Weinstein

10   9   8   7   6   5   4   3   2   1

*Cover photograph: Power plant coal pulverizer parts. Orange objects in foreground
are grinding balls made of special metal each weighing 254 pounds.*
Photo courtesy The Babcock & Wilcox Company.

This book is dedicated to that unique group of clients who saw a better way to do something and embraced it. You are too numerous to mention individually, but you know who you are.

TABLE OF CONTENTS

CHAPTER 9: PARTNERING WITH OTHERS

CHAPTER 10: SOME BEST PRACTICES AND LESSONS LEARNED

CHAPTER 11: IMPLEMENTATION

PREFACE

This book is about managing production spare parts-the materials that keep your plant equipment running when something fails. Most capital-intensive industries such as chemical, pulp and paper, refining, electric generation, and mining have warehouses with millions of dollars of insurance spares devoted to keeping their production equipment operational. Although these parts may be infrequently used, uninterrupted and safe production depends on their availability when needed. Worldwide competition for business today has encouraged a more comprehensive look at spare parts inventories and how they are managed. Top-level industrial companies that have taken advantage of modern inventory management strategies and tools have enjoyed beneficial transformations in the size and composition of their inventories. There is no reason why any process-intensive company can not reap those benefits as well.

Having a well-managed spare parts inventory means having what you need when you need it-not too much, nor too little in the storeroom. It's been said that "you can't predict the unpredictable." That's true. But this doesn't mean you can't manage around unpredictability. Predicting when things are going to fail has always been an unrewarding task. There's the new light bulb that fails as soon as you first turn it on, despite the 2,000-hour guarantee, or the battery that never seems to run out. We could find many more examples.

Techniques for trying to predict spare parts failure tend to take into account the expected frequency of failure. Parts that fail often, even if not in a uniform pattern, are more likely to be predictable using today's modern mathematical models and algorithms. Parts that fail infrequently (e.g. several years between failures) are not. For these items, more sophisticated techniques such as lumpy demand (sometimes called stuttering Poisson) have been tried with limited success, as long as there have been at least a few demands for the part over the last couple of years. In this book, we will introduce another concept, risk-based assessment.

We will also distinguish between active and infrequent (rarely-used) demand. The usage rate of a part has been found to be the most important factor in determining which inventory management techniques should be applied. We have chosen a cutoff that defines an active item as one whose average usage rate is more than one per month (more than 12 per year); conversely, a rarely-used item has one demand per month or less.

There are two reasons for setting the cutoff at this level: 1) even in the more sophisticated multi-model forecasting systems, some models fail to work properly below this rate of usage, and 2) most inventory items are issued either significantly more than once per month, or are hardly used at all. As we will show later, many good forecasting models exist for active items; these models can use historical monthly demand to predict future independent demand with fair accuracy. Commonly-used models include variations of linear regression, exponential smoothing, seasonality, and moving averages. Although forecasters that are properly applied work very well for active items, the vast majority of maintenance, repair, and operating (MRO) inventory is too infrequently used to allow these techniques to be applied.

Contrary to the attributes of actively-used inventory, rarely-used items are characterized by the following: 1) they are issued infrequently (typically 40-to-60 percent of rarely-used items at a site have had no usage in the last three years), 2) they are non-forecastable using traditional techniques, 3) they have high cost per unit, 4) they have high inventory values, 5) they have longer lead times, and 6) they tend to be more critical to supporting operations. As we will show later, the concept of service level does not apply to rarely-used items.

Another small group of stores items, called commodity items, cannot be ignored. Commodities tend to have a variable quantity of issue. They constitute only a small amount of the total plant inventory and are usually not critical to operations. They have short lead times and often fluctuate widely in demand. Because most of the inventory at plant storerooms is slow moving, most of the material in this text will focus on that topic.

One thing we will stress throughout this text is the need to focus on what's important when managing the plant spare parts inventory. This need brings us to Pareto's rule (the 80/20 rule). Because resources to manage inventory are limited, spares need to be sorted into key and non-key items. This step is important because it will determine the number of items (and the amount of effort) that need intensive management (the key items). In turn, non-key items can be managed when, and if, time and personnel resource permits. Pareto's rule should also apply when monitoring and reporting on inventory status so that all reporting is in order-of-priority. The top 10--20 per cent of the items will have about 80 percent of the impact.

Significant corporate assets are tied up in spare parts inventories. For example, a nuclear power plant can easily have $50-to-$100 million in inventory at a single site. Work by the authors at over seven hundred plants worldwide has shown that, depending on industry, 25-to-50 percent of the inventory investment is not necessary, but is excess to the needs at the plant to meet availability objectives. Once overstocks are identified, they can be used or disposed of to free up cash for more productive uses.

In choosing the content for this book, we have structured the material to reflect not only the differing interests at a plant for managing spares inventory but also how each can impact the supply chain. For example, purchasing managers and buyers will understand the importance of procurement lead time and how shorter lead times can mean less inventory investment. Maintenance personnel will have a better understanding of the risks involved in stocking spares; they will learn when enough is enough. Storeroom personnel will understand how a spare part can impact production and affect the number of spares required. Plant management will be able to set reasonable, measurable goals for controlling the inventory asset.

The text of this book is structured into eleven chapters. Chapter 1 puts the spares inventory into perspective in relation to other company assets, especially in the manufacturing process. Characteristics of active and rarely-used inventory are defined, as are their typical plant inventory profiles. Some standby techniques, like the A-B-C analysis, are discussed along with a new concept we call the inventory tree. Chapter 2 addresses the important subject of risk. We will show that a cost bias exists in virtually every production plant when stocking spares. As a result of this bias, inexpensive spares are grossly overstocked and expensive spares are normally understocked to meet availability objectives. The problems of trying to predict future demand will be covered as well as how to use sensitivity analyses to overcome uncertainty when stocking parameters vary widely. We will also look at some risk assessment case studies.

Chapter 3 will discuss the process of setting the reorder point: the stock level at or below which the replenishment process should begin for both active and rarely-used items. We will also cover another common bias when stocking spares inventory, the lead time bias. Most important, we will show how setting the reorder point is sensitive to various input parameters such as criticality, lead time, and demand. In turn, Chapter 4 focuses on setting the economic order quantity (EOQ), the amount that should be bought when an order is placed. The chapter covers how the EOQ is impacted by price discounts along with different concepts for managing the replenishment process.

Chapter 5 highlights ways to determine how much of the plant inventory is gross excess, using a novel concept called the extreme test. Once overstocks are identified, ways to avoid the problem are discussed with the objective of achieving a balanced inventory. Then in Chapter 6, a concept for avoiding unnecessary purchases is covered, along with some new approaches for buying new spares and reducing replenishment lead times. A number of unique solutions for fixing some everyday problems are the main focus of Chapter 7, whereas Chapter 8 discusses the process of setting and monitoring goals. The much misunderstood subject of turnover is also covered. We will also introduce what we believe to be the most important measure of inventory balance, the Absolute Variance Ratio, or AVR. The discussion considers how AVR and the so-called clothes line chart can be a vital tool in monitoring inventory management performance.

Chapter 9 covers the timely topic of partnering with others such as your supplier or friendly consultant. Chapter 10 focuses on what the authors believe to be some of the best practices for managing active and rarely-used inventory, and some of the important lessons learned from our nearly seventy years of collective experience in consulting on spare parts inventory. Finally, Chapter 11 looks at implementation, discussing ways to make each inventory project more beneficial. We have made great efforts throughout the book to keep it from becoming overly commercial. We have also used the terms reorder point and MIN interchangeably, and they should be considered to have the same meaning.

Case studies and examples used in the book are based on real client data, or adaptations thereof, but with stock numbers, part descriptions, prices, etc. changed to protect the security of client data. In most case studies, and there are forty-eight, the situation depicted has been simulated. When a client's name is used it is with their permission, as is also true with certain supporting material from published sources. When formulas or excerpts from other published works are used, credit is always given in the list if figure credits.

Material on various material management systems or software packages for managing or stocking inventory was obtained from published sources, such as web sites, or from direct discussions with personnel from the firms. All case studies, while using real data, are named for a fictional company call AJAX, and any resemblance to a real company is purely coincidental. Any opinions that are stated are strictly those of the authors.

The authors wish to thank Leon Daggett, Gary Haggerty and Tony Pike of the Kansas City, Kansas, Board of Public Utilities; they have piloted all of the concepts discussed in this book, and have proven that, by implementing them properly, it is possible to obtain a near-perfect inventory balance.

Our thanks are extended to the following who assisted by reviewing the material in Chapter 9 on partnering with others: John Dinyari, Kurt Mitchell (Scientech), Alan Skedd (Decision Associates Inc.), Lyra Ward (Indus), Jim Couch (IHS- Intermat), and Roger Hutton (spares-Finder). Some additional reviewers elected not to be cited.

We appreciate the help of Barry Rippe (NiSource) and Steve Bryk (The Babcock and Wilcox Company) for providing the spare part photos used throughout the book.

Appreciation also goes to colleagues, past and present, who helped develop many of the concepts covered, reviewed the manuscript, or provided advice, especially Alan Skedd, David Hamilton, and Kim Johnson. We also wish to thank the numerous client program managers who worked with us over the years and helped implement many successful projects.

We were told by some when we started this book that working with a publisher can be a difficult experience. Not true in our case. It was a real pleasure working with the team at Industrial Press, including Patrick Hansard (Sales and Marketing Director), John Carleo (Editorial Director), Janet Romano (Production Manger/Art Director), and Robert Weinstein (Manuscript Editor), a true wizard with words.

LIST OF FIGURES CEDITS

LIST OF PHOTO CREDITS

All below mentioned are Courtesy of The Babcock & Wilcox Company

On front cover	Power Plant Coal Pulverizer Parts
On back cover	Tubing and roll wheel tires
On back cover	Roll wheel brackets
Page 96	Tubing
Page 106	Spare part racks

All below mentioned are Courtesy NiSource

Page 40	Bushing
Page 45	150 hp motor
Page 84	Plug valve
Page 158	Circuit breaker
Page 166	Valve guides
Page 169	Manway gaskets
Page 184	5 hp motors
Page 197	Mechanical wear blocks
Page 236	Mechanical elbow
Page 286	Centrifugal Pump

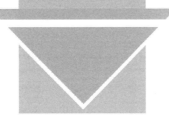

INDEX OF CASE STUDIES

CHAPTER 1

INVENTORY AS AN ASSET

1.1   WHAT THE READER SHOULD LEARN FROM THIS CHAPTER

- How inventory fits in determining Return-On-Iinvestment (ROI)
- Types of inventory in manufacturing
- Key production inventory issues
- The difference between active and Rarely-used inventory
- Typical inventory profiles
- The inventory tree concept
- How an A-B-C analysis works
- Three different inventory management techniques

1.2   RELATIONSHIP OF FACTORS AFFECTING ROI

Accounting practices among companies vary widely regarding what are considered to be inventory, capital, and expenses. Therefore, values at an individual company may be different from those used in this book, which are an average of companies with differing practices. Readers should keep these differences in mind during the discussion of ROI in this section, and throughout the book when discussing inputs for determining stocking levels.

It's been said that the best way to get to the core of something is to peel it apart in layers like an onion. This is true when determining where inventory fits among the overall elements that make up the return-on-investment calculation, also known as return on assets (ROA)

Consolidated Balance Sheet (in millions)	*AJAX Manufacturing*	
December 31	2003	2004
Assets		
Current Assets:		
Cash and cash equivalents	34	51
Short-term investments	6	2
Receivables from customers	126	130
Inventories	231	212
Deferred income taxes	19	23
Prepaid expenses and other current assets	31	34
Total Current Assets	447	452
Property, plant, and equipment, net	482	514
Goodwill	41	44
Other Assets	206	211
Total Assets	1,176	1,221

Figure 1.1 Balance Sheet and Operating Statement

and return on net-assets (RONA). Figures 1.1 and 1.2 show the many factors that make up a typical balance sheet and operating statement for a company. Notice that current values for cash, receivables, inventory and property, plant and equipment are on the balance sheet, whereas sales, general administrative expenses, and operating income are on the statement of consolidated income. Both the balance sheet and the operating statement are valid only for the day stated, in this case December 31.

Figure 1.3 shows how these factors are related in determining ROI. Once we calculate Turnover and the Profit As % Of Sales, we simply multiply the two to find the return on investment (14.6 percent for AJAX Manufacturing). We can improve ROI by increasing turnover, which means either increasing sales or lowering total assets, including inventory. We can also improve ROI by increasing operating income on sales, which usually means cutting labor, material, and factory overhead costs.

Statement of Consolidated Income (in millions) For the year ending December 31	*AJAX MANUFACTURING* 2003	2004
Sales	589	642
Cost of goods sold	202	240
Selling, general administrative expenses	76	84
Research and development expenses	23	27
Provision for depreciation, depletion	82	72
Interest expense	39	42
Income from continuing operations before taxes	167	177

Figure 1.2 Statement of Consolidated Income

Financial managers frequently believe that inventory is too mundane a part of the ROI calculation to deserve much attention, at least compared to labor, material cost, and factory overhead. For this reason, few resources are employed to manage inventory; seldom are inventory metrics reported routinely to top management, especially compared to reporting on sales figures. Because inventory gets little attention from top management, relatively few people set or monitor inventory goals in the organization. In the subsequent chapters of this book, we will show that by employing some practical techniques, we can optimize the amount of inventory needed to meet plant operating objectives and also provide some useful metrics for setting and moni-

Continuing from Figures 1.1 and 1.2, the return on investment can be calculated as follows for 2004:

TURNOVER   = **Sales divided by total assets**

PROFIT AS % OF SALES = **Operating income divided by sales**

TURNOVER 0.53	times	PROFIT AS % OF SALES 27.6	=	RETURN ON INVESTMENT 14.6 %

Figure 1.3 Calculating the ROI

toring inventory goals. The end result will be a lower inventory asset and an improved ROI. Now that's a noble goal.

1.3   TYPES OF INVENTORY IN MANUFACTURING

All inventory is not alike. For example, retail inventory includes clothing, TV sets, grocery stock, etc., whereas production inventory may include steel plate and spare parts that are vital to maintaining the manufacturing process. It is production inventory that is the focus of this book, especially spare parts.

The need for inventories is created by management's desire to operate the various purchasing, manufacturing, distribution and sales operations somewhat independently of each other. Even with what appears to be perfect planning, supply chain disruptions will occur, affecting the smooth flow of production if there is no inventory cushion (Murphy's Law). Inventory concepts like just-in-time can work, but they frequently require the facility supplying the parts or components to be adjacent to the assembly operation to avoid transportation disruptions.

Inventory in a manufacturing process is often broken down into four categories: 1) Finished Goods, 2) Work-in-Process, 3) Purchased Raw Materials, and 4) Operating Supplies and Replacements Parts.

Inventory breakdowns can be found on the Notes to Financial Statements:

Notes to Financial Statements D. Inventories December 31	*AJAX Manufacturing* (million dollars)	
	2003	2004
Finished goods	143	136
Work in progress	56	39
Purchased raw materials	21	24
Operating supplies, replacement parts	11	13
	231	212

Figure 1.4 Types of Inventory in Manufacturing

To illustrate, we have excerpted a Notes to the Financial Statements for AJAX Manufacturing (see Figure 1.4) taken from their annual report. It is primarily from the last line of the Notes that we first encounter any comment on spare parts.

Figure 1.5 shows examples of each type of inventory. Any factory that manufactures a good using production machinery will have the need, from time-to-time, for replacement spare parts. A decision must be made whether to stock the spare as an individual part or as an assembly. Frequently it is faster, and less costly, to replace an entire component of a machine than try to search out the failed part and hope a replacement is in the storeroom.

1.4   KEY INVENTORY ISSUES

Several inventory issues frequently require attention. We have encountered all of these in our work with clients around the world.

1.4.1 Eliminating Excess Inventory

There is a natural human resistance to change: people are comfort-

The need for inventories is caused by management's desire to operate the various purchasing, manufacturing, distribution, and sales organizations somewhat independent of each other

Inventory type	Examples
Finished goods	Pumps Automobiles Packaged boilers
Work in progress	Sub-assemblies Reworked production
Purchased raw material	Steel plate Crude oil
Supplies, spare parts	Batteries Rope, dope, soap

Figure 1.5 Types of Inventory in Manufacturing

able with what they already know and understand. A change in inventory management philosophy will likely invoke some uneasiness among material managers and operations staff. Chief among their concerns is "You're trying to take away my inventory (and threaten my parts availability)" and "This is going to create a lot of additional work (for me)." The focus should not be just on inventory reduction, but also on inventory balance. When approached properly, these can be achieved with a minimum of effort and risk to operations.

Historically, stock levels for individual items have been determined manually by a combination of inputs from vendors, engineers, and maintenance. Experience has proven that these well-intended best estimates come with inherent biases that are both conflicting and counterproductive. Usually the process starts like this: someone calls up maintenance and tells them that for some new part we must stock a minimum and maximum value in the material management system that determines when to reorder the part and how many to buy. Then maintenance says "set the MIN at 2 and the MAX at 6" because they believe those are reasonable and appropriate values. What they may not understand is that using 2 and 6 for the MIN/MAX now sets the re-order quantity at 4 units even though the economic order quantity (EOQ) formula suggests buying only one unit. (See Chapter 4 for more on the EOQ.) Furthermore, if the part ever stocks out and causes any grief for maintenance, the odds are they will raise the stocking level by a significant amount to make sure such an event never happens again.

Vendor recommendations cannot always be trusted either even though vendors should have a fairly good idea how often parts in the equipment they sold are likely to fail. Our experience over the last twenty years is that vendors recommend too much inventory about 70 per cent of the time, and not enough about 10 per cent of the time. Of course, recommending more than is needed also happens to be profitable to the vendor on the sale. Consider a recent cartoon showing two clerks behind the counter of an auto parts dealer. One says to the other, "You've got to hand it to the automakers-they can take $70,000 of replacement parts and make a $20,000 car out of them."

From analyzing inventory levels at hundreds of production plants around the world, we also know that between 20 and 50 percent of the value of typical storeroom inventory is excess to the plant's needs. Identifying and reducing the amount of excess inventory is clearly important. Knowing how can be the hard part.

1.4.2 Setting Initial Spares Levels

Setting initial spares levels can be very difficult because there is no previous usage history on which to rely. As noted before, depending entirely on the vendor' recommendation can be costly. If prior usage history is not available, then some estimate of the expected mean-time-before-failure (MTBF) for the part is needed in order to estimate initial requirements. Our experience is that vendors are reluctant to supply MTBF information. Their attorneys fear that if the customer uses the data to set initial spares and the part fails sooner than projected, the customer will sue for lost production damages (consequential damages). Therefore, don't count on getting MTBF data from your vendor. You will see later in this book a more effective way to set initial spares.

1.4.3 Improving Availability

Without question one of the biggest concerns of plant maintenance personnel is that their supply of replacement spare parts is not large enough to prevent loss of production due to lack of a spare. Prior to the development of risk-based techniques, a variety of methods were used to attempt to manage rarely-used spares. Asking a vendor how much to stock has proven to result in 20-to-40 percent overstocks. Asking maintenance personnel results in 25-to-50 percent overstocks with several biases including a cost bias and a lead-time bias. Asking the materials manager to guesstimate the required stock commonly results in understocks compounded by the same biases.

When setting spares stocking levels, maintenance personnel should be asked only two questions, both of which they are eminently qualified to answer. The first question is, "If you need one of these parts and one is not available in the storeroom, how much will it impact production?" In other words, what is the part's criticality? The second question is, "When you install one of these parts, how long do you expect it to last?" Simply stated, estimate the MTBF for the item (provided reliability values are not available from the supplier, which they seldom are). One question not to ask is, "How many do you think we should buy?" Later, we shall cover how to set criticality for a part. We will also consider the other input parameters that go into the decision of how many to stock in order to protect availability without over- or understocking.

One final point about availability. For years, television viewers have been amused by the lack of service calls to the friendly Maytag repairman. This is the classic "good news, bad news" story. It is good news for the owners of the washer because they bought a quality product which is working just fine. This is also good news for Maytag because it enhances their image as a high-quality producer. The bad news applies to the Maytag service organization because they are not selling high-margin spare parts. The point of the story is this: maintenance people would love to be like the Maytag repairman-no breakdowns, no need for spares, no lost production, just lots of time for coffee breaks! And don't forget petting the dog.

1.4.4 Setting And Monitoring Goals

Top management , and sometimes regulatory agencies, frequently want to know how well plant inventory is being managed. This request is reasonable because spare parts inventory represents a sizable asset for most production plants. Inventory turnover ratios are often quoted, although we don't think this metric is an appropriate one, as we will discuss later in this book. Suffice it to say, there are only a few very good measures that allow management to keep their finger on the pulse of the inventory asset and to determine if steady progress is being made toward better managing the inventory asset. What are some of these valid measures of inventory performance? We suggest that three will suffice: the absolute variance ratio (AVR), the +/-1 percentage, and the net change in actual inventory. We will cover all of these in Chapter 8 in more detail and with some case studies.

1.4.5 Investment Recovery

Identifying overstocks is more than finding items with a quantity-on-hand greater than the current maximum. Typically, initial minimums and maximums have been determined by vendor recommendations. More often than not, this quantity has been too high. Over time, a few maximums may have been reduced. However, if the storeroom ever was caught short, the maximum for that item has probably been raised to a quantity that more than accommodates the "worst ever" condition.

Attempts to write off or otherwise dispose of excess inventory are usually received by operating personnel with the comment, "Why dispose of it since we already paid for it, and who knows we may need it again sometime?" Although investment recovery is the response, most companies stop at the excuse, "But I don't have the budget for write-offs." Thus, the failure to act in the overall best interest of the company burdens operations with substantial overstocks and costs that can and should be avoided. Investment recovery will be addressed in more detail in Chapter 5.

1.5   TYPES OF PRODUCTION INVENTORY *

We like to distinguish between two types of inventory: active and rarely used. Another type of inventory, called commodities, can be either active or rarely used.

1.5.1 Active Inventory

Active items are used frequently enough (e.g., more than once per month on average) that future expected demand can be predicted with good accuracy. Active items: 1) are usually commodity or smaller

---

*Much of the material in the remainder of this chapter is based on *Spare Parts Inventory Management*, by Michael P. Reynolds, APICS, April 1994.

spare parts, 2) have generally high demand each month (e.g., 100+ per month), 3) have predictable future demand, and 4) are frequently seasonal. These parts need an active item forecaster to predict future demand.

For active items, the priority objectives are to improve service level, reduce item cost, and minimize transactions cost.

Improve Service Levels

If a fast food restaurant runs out of french fries, you would have a poor opinion of its operations. Suppose this happened on a family vacation years ago. The kids would still remember the name of the city without french fries, even though they can't remember the names of the four faces on Mt. Rushmore, which was visited the next day (left to right Washington, Jefferson, Teddy Roosevelt, Lincoln).

Similarly, if your storeroom runs out of work gloves, flashlight batteries, or standard hex-head cap screws, your customers would be justifiably dissatisfied and might long remember. Active items account for the majority of all potential customer satisfaction events: requests for stock. The primary benefit of multi-model forecasting software tools is improved service levels, not inventory reduction. If salespeople try to justify your investment in a forecasting tool based on lowering inventories, throw them out because they not only do not understand your business, they do not even understand *their own* business.

Reduce Item Costs

The repetitive purchase of active items can increase the cumulative total value of the savings derived from negotiating lower unit cost per item. Although the active items are a small percentage of the total items, they can account for almost half of the total usage value. Furthermore, any savings, such as a 10 percent per unit cost reduction, is repeated over and over each and every year.

Minimize Transaction Costs

Transactions include the activities of picking, delivering, tracking on-hand quantity, communicating with vendors, receiving, incoming quality control, and paying accounts payable. By definition, the majority of transactions are for the active items. Transaction costs can be reduced by implementing programs such as integrated supply chains, strategic vendor alliances, electronic data interchange, vendor quality certification, consignment, rack jobbing, floor stock, and credit cards.

Reducing the inventory investment in active items is a relatively low priority. Active items comprise a small percentage of the total items and an even lower percentage of the total inventory value. The more active the item, the stronger are the tendencies to have good forecastable demand, short lead times, low unit costs, low risk of obsolescence, and low likelihood of exceeding shelf life.

The total inventory required to support high service levels for active items is a small fraction of the total inventory. Therefore, reasonable expectations for never being out of stock should be very high because the implications of carrying some surplus are not very serious. Thus, the objectives for active items ought to be to implement programs which improve service levels, reduce item costs, and minimize transaction cost.

Active items exhibit these characteristics:
- They are typically used in large amounts.
- They have a high frequency of issues.
- They tend to be the less expensive items.
- They have fairly short lead times to replenish.
- They are seldom critical to maintaining production.
- They need high service levels.
- They can be forecasted.
- Because they move fast, overstocks can correct quickly.

In most production operations, active items make up only about 10 to 15 percent of the total storeroom inventory.

1.5.2 Rarely-Used Inventory

Rarely-used items are used too infrequently (e.g., once per month or less) to be able to predict future demand with reasonable accuracy.

The vast majority of the items in a typical operating and maintenance storeroom are rarely used (an average of 88 percent in hundreds of storerooms we've reviewed). About half of these items have had no usage whatsoever during the prior two years, yet most of them must be on hand when needed or the consequences are dire.

Three statements of fact about rarely-used items: 1) assume you have no idea when a rarely-used part will be needed because usually you don't, 2) no amount of inventory investment will guarantee there will never be a backorder of any item under all possible conditions, and 3) no amount of inventory reduction can save enough money to compensate for the lack of a needed highly-critical spare.

Whereas high service levels for active items can be achieved by using a multi-model forecaster to plan for seasonal, increasing, or decreasing demand patterns, the future demand pattern for a rarely-used item cannot be forecasted. Because forecasting techniques do not work, rarely-used items require a risk-based tool to minimize total costs.

For example, an auto insurance company may have a reasonably accurate forecast about how many claims will be incurred next Friday, but it doesn't have a clue about who will file them. If they did, they would probably cancel those people's insurance next Thursday. Much the same can be said about rarely-used inventory; we know there will be some items needed next week, but we have almost no idea which specific items. If we could know which items will fail next Friday, we would have ordered them to arrive by next Thursday. Just-in-time is an irrelevant concept for rarely-used items because the timing of future demand is unknown. Instead, Just-in-case is the more appropriate term.

More inventory will not eliminate all backorders, yet less inventory cannot save enough money to compensate for dire consequences. Therefore, the prudent business objective is to minimize the total cost of ownership and non-ownership of each rarely-used item. The priority objective then, for rarely-used items, is to minimize the total risk cost: the sum of the costs of getting caught long (carrying inventory when not needed) plus the costs of getting caught short (incurring backorder costs when needed and out of stock).

Why will we focus predominately on rarely-used inventory in this book? For three reasons: First, it is the inventory that is most important to maintaining production. Typically 80-90 percent of production spares are slow-moving and represent most of the inventory value. Second, it represents the greatest opportunity for improvement. Finally, there is relatively low investment to implement improvements. Using Pareto's Rule (the 80/20 rule) can significantly reduce the personnel and financial resources needed to implement improvement. In short, spend your time and effort where the benefits are!

Rarely-used items exhibit these characteristics:

- They are issued infrequently (40-60 percent have no demand over the last three years).
- They can not be forecasted using traditional techniques.
- They typically have high cost.
- They make up most of the inventory value at plants.
- They have long lead times (frequently a year or more).
- They tend to be critical to maintaining production.
- They need extremely high availabilities at the piece-part level to support production.

1.5.3 Commodity Inventory

The management of commodities is a low priority because they represent such a small percentage of inventory value. Furthermore, no single forecasting technique will successfully work for commodities all the time. Like active items, most commodities have short lead times so that understocks have the potential to be resolved quickly. Severely overstocked commodities can run the risk of shelf life expiration or part obsolescence. Because most are inexpensive, however, the loss is minimal.

Commodity items exhibit these characteristics:

- They tend to be issued in variable amounts (by the pound, gallon).
- They are usually less than 5 percent of production inventory.
- They are seldom ever critical to maintaining production.
- Substitutes are readily available through local sources.
- Usage often fluctuates widely, and even seasonally.
- They may have a finite shelf life.
- They can be difficult to forecast.

1.5.4 Optimum Inventory Level

Is there such a thing as an *optimum* inventory level? The answer is "yes," as you will see in Chapter 10 where one electric utility actually came within 3 percent of achieving it.

Understanding the optimum level can be simplified by starting with a look at the absurd extremes. On the low end, what if no inventory of any item were stocked? Every time a part was needed, purchasing would call the vendor and try to expedite a replacement. The Federal Express and UPS trucks would be swarming around the receiving dock, expedite costs would be horrendous, and production would be almost non-existent because even the most routine repairs would become scenes from a Three Stooges movie. The backorder costs, primarily lost revenues from low production plus the expedite premiums, would be exceptionally high.

On the high inventory end, what if absurdly large quantities of every item were stocked? For example, what if a valve that was used maybe once per year, with a lead time of four weeks, were stocked at 100 units and was reordered every time one was used. Under even the most unlikely conditions, the on-hand stock will rarely be lower than 98 units. The carrying cost resulting from such an absurd policy for every item would be exceptionally high.

Between these two absurd extremes is a range of more sensible stock levels with lower total backorder plus carry costs. Some of these alternatives have lower cost implications than others. Using a risk-based decision support tool, it is possible to determine each individual item's optimum stock level with the lowest total cost. Summing the optimum stock for each item then results in the optimum total inventory level for all items.

Many companies have installed new computer hardware, online or real time material systems, common sense rule-of-thumb guidelines, and complex mathematical algorithms. Despite all of these, the result is inevitably the same-continued increases in inventory and a degradation in the availability of items for operations and maintenance. But there is hope for improvement, as you will find out later in this book.

1.6 TYPICAL PRODUCTION INVENTORY PROFILES

Figure 1.6 shows typical MRO (maintenance, repair and operating) inventory profiles. Looking at the two left bars, you can see that rarely-

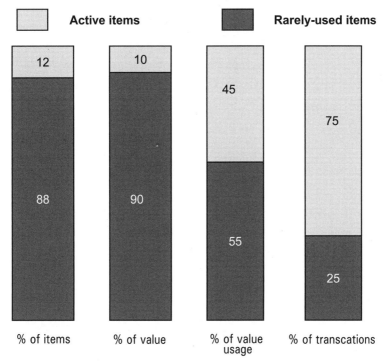

Figure 1.6 Typical MRO Inventory Profiles

used items make up about 88 and 90 percent of total storeroom items and inventory value, respectively. These high levels are consistent across most industrial segments such as power generation, petroleum refining, chemical production, steel making, and paper production. Looking at the right bar, notice that active items account for up to three-fourths of the stock transactions. The job of the MRO materials professional is to employ innovative strategies for optimizing each of these segments of inventory. Years of experience has shown that when it comes to inventory, one strategy does not fit all! What works for active items generally does not work for rarely-used items. We will discuss this aspect in greater detail in Chapters 2, 9, and 10 of this book.

1.7   THE INVENTORY TREE

One way to sort out the best strategy for optimizing both active and rarely-used inventory is to adopt an approach called the "inventory tree" (see Figure 1.7). This approach was developed to manage spares in a more efficient and beneficial manner. Basically, this philosophy involves classifying inventory according to its characteristics, then applying the appropriate resources to manage each inventory class relative to those characteristics and the item's importance to production.

We have found several advantages to this approach. First, it is uneconomical or inefficient to manage all inventory aggressively. By categorizing inventories and applying different management techniques to each, it's possible to maximize the value of the stocking decision with a minimum of effort. Second, different tools for managing inventory are applicable to different inventory classes. The stocking logic for an actively-used, routinely-ordered item is quite different from that of a rarely-used, highly-critical one. Third, categorizing inventory will permit materials managers to spend the most amount of time improving stocking decisions of the most important spare parts, rather than giving equal attention to any item that crosses their desk. Finally, the inventory tree sets quantifiable criteria for each inventory classification.

Figure 1.7 shows how the inventory tree ideology was applied to in-

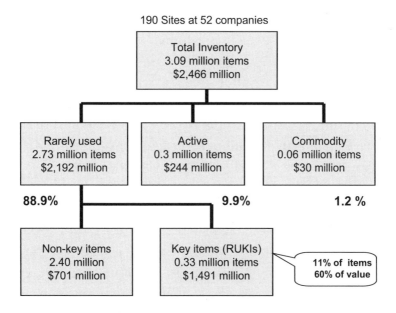

Figure 1.7 The Inventory Tree

ventory at 190 sites for 52 companies in the power, pulp and paper, chemical and petrochemical, refining, and railroad industries. Total inventory value was $2.4 billion consisting of 3.1 million stock keeping units (SKUs). It was possible to group these industries together because the results throughout these industries are similar, indicating that spare parts inventory management is not so much a function of industry type as it is the characteristics of the spare parts themselves as listed above.

The first step in the analysis is to separate the spares into active and rarely-used classes based upon usage rate. A part's usage rate has been found to be the most important factor in determining which inventory management technique should be applied. We have elected to define a rarely-used item as any item that is used an average of once per month or less (90.0 percent of the total inventory in Figure 1.7). In the process industries, active items comprise roughly 10 percent of the items, the same as shown in the inventory tree example. The next step is to separate the rarely-used (non-commodity) items into key and non-key groups. This step is important because it determines the number of items (and the amount of effort needed) which justify intensive management (the key items) as opposed to those that can be managed when convenient. It is also an opportunity to see the Pareto rule in action, where focusing on a small percentage of inventory line items will effectively address the majority of the inventory value.

To identify these rarely-used key items (hereafter referred to as RUKIs), a decision must be made on a key item cutoff-a measure of inventory value or periodic usage-such that if an item exceeds either condition it is considered key to operations. Conversely, parts which fail both tests are non-key items. As the key items cutoff is increased, the number of RUKIs decreases. We usually try to find the cutoff that captures 85 percent of the rarely-used inventory value; this level nor-

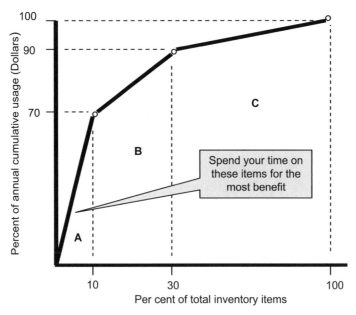

Figure 1.8 Typical A-B-C Analysis

mally requires looking at only about 10-15 percent of the stock items (11 percent in our inventory tree example). Several things can be said about RUKIs:

- They have a disproportionately larger share of inventory value relative to the number of items.
- Managing only these items results in the greatest benefit for the least investment.
- Nearly one-half of all RUKIs tend to be critical to operations.
- They exhibit a cost bias, which we will discuss in more detail in Chapter 2.

Two final thoughts on managing RUKI inventory. First, because RUKIs are used infrequently, essential information like lead time and price are not updated very often from purchases, thereby making this information of limited use to material management personnel. Second, because of their limited usage, setting a stocking level too high can tie up cash for extended periods, instead of it being applied elsewhere.

1.8   THE A-B-C- ANALYSIS

The previous discussion about RUKIs is a natural lead-in to looking at the typical A-B-C analysis. As we have said, a few items normally account for most of the inventory and usage of spare parts. This is best represented in Figure 1.8, which plots cumulative usage against total inventory items. Group A items are those items that comprise about 70 percent of the inventory value, but only 10% of the line items; Group B, 20 percent of the value and about 20 percent of the items; and Group C, only 10 percent of the value, but all of the remaining items.

Clearly the Group A items are the RUKIs. It is the management of these items that should be the highest priority for the material management team. Several suggestions for doing that are: 1) limit management's effort so that it works only on the RUKIs, 2) focus only on those RUKIs that are out-of-balance, that is, RUKIs that have reorder points or reorder quantities higher or lower than optimum, and 3) prioritize reporting of out-of-balance RUKIs so that the "big hitters" are the first to be reviewed and adjusted, and the "small potatoes" are reviewed last, if at all.

1.9   INVENTORY MANAGEMENT TECHNIQUES

Over the last years several different techniques have been devised to improve the availability of spare parts when needed. Let's look at the three best:

1.9.1 Material Requirements Planning (MRP)

The philosophy of MRP is based on the concept of determining the future time-phased requirements for material and taking the necessary actions to insure the material is on-hand when required. Critical path scheduling is often used to keep the project on schedule. The key to successful MRP is the scheduling process, which determines when certain activities are going to happen. For example, if an overhaul of a utility turbine generator is scheduled for the April spring outage, ordering of expected materials for the overhaul must be backed dated from April and reflect the expected lead time to manufacture or procure the materials. Besides overhauls, other activities that frequently use MRP are assembly line operations and equipment preventative maintenance procedures.

1.9.2 Distribution Requirements Planning (DRP)

DRP is normally used in free issue situations where stocks such as fasteners, screws, and flashlight batteries are available for maintenance to pick up without filling out an issue ticket (just help yourself). Generally the stock items are low cost and purchased in bulk so that precise inventory control is not justified.

1.9.3   Non-Forecastable Requirements Planning (NRP)

NRP is the process used when you have no idea when the material will be needed, if at all, and you must rely on the use of risk-based processes to set stocking levels that will meet operations availability

What we DO and DON'T know		

Technique	Know how many are needed	Know when they are needed	Know why they are needed
MRP Material Requirements Planning	YES	YES	YES
DRP Distribution Requirements Planning	YES	YES	NO
NRP Non-Forecastable Requirements Planning	YES	NO	YES

Figure 1.9 Inventory Management Techniques

objectives. The principal use of NRP is the management of rarely-used items; MRP and DRP won't work here.

Figure 1.9 summarizes what we know and don't know about a situation when employing MRP, DRP and NRP techniques. For all three techniques, we usually have a pretty good idea of how many of some item will be needed. With MRP and DRP, we usually know when it will be needed, but not with NRP. And with MRP and NRP, we usually know why we need something; with DRP, we are not concerned with why. Because each technique has its advantages and disadvantages, the key is to apply the techniques where it works best.

CHAPTER 2

ASSESSING RISK

2.1 WHAT THE READER SHOULD LEARN FROM THIS CHAPTER:

- How Spare Parts Typically Fail
- Some Statistical Distributions that Impact the Risk of Failure
- Some Approaches for Setting Stock Levels for Slow-Movers
- Ways To Predict Future Demand
- Some Models for Forecasting Active Inventory
- Understanding Service Level
- How To Assess Risk for Rarely-Used Items
- How Sensitivity Analysis Can Be Used To Reduce Risk

2.2 HOW SPARE PARTS TYPICALLY FAIL

This chapter is about equipment parts that fail in service. If we were fortune tellers, we could predict exactly when something was likely to fail and then order a replacement to arrive before the part did fail. However, no such luck! Parts tend to fail when they want, not when we want. That's why we have safety spares to have a replacement on hand in case of unplanned failures.

This perspective provides a good lead for the following definitions of probability and risk:

Probability is the likelihood that something will happen and risk is the likelihood that something bad will happen.

Although we can't predict precisely when something is likely to fail, we are able to understand failure rate patterns and plan accordingly. Figure 2.1 shows a typical failure rate pattern. Notice the high incidence of early failures where parts tend to fail when first exposed to the rigors of service. Examples include light bulbs that blow as soon as you turn them on, circuit boards that fail as soon as the computer is stated, and motors that short out the first time they are started.

Usually such failures are due to quality problems not caught during manufacture. Nevertheless, they contribute to the normal failure pattern. As time passes, parts tend to break in. Failures then decrease for an extended period, gradually increasing as the equipment ages. We will show in Chapters 3 and 4 that while it's a good idea to carry extra spares early to cover startup spikes, it not advisable to include these failed spares when trying to predict future demand after steady-state conditions are reached.

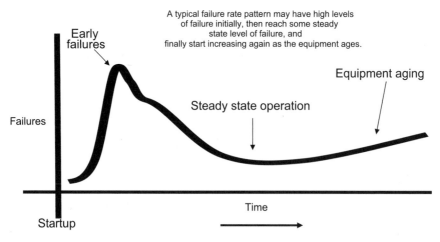

Figure 2.1 How Parts Typically Fail

2.3 PROTECTING PLANT AVAILABILITY

Because we haven't been able to predict random failures of parts in service, we must resort to the next best approach stocking-backup or safety spares. Deciding how many to stock is what most of this book is about. The stocking decision depends on many factors including: 1) whether the demand for the part is very high or very low, 2) how important the part is to maintaining production, 3) how long it takes to replenish the storeroom stock, and 4) how many parts are used in production.

Figure 2.2 illustrates the complexity of the problem. This chart shows the impact of one of four parts in service failing (each carrying 25 percent of capacity) with no spares available in the storeroom. Initially, production is rolling along at 100 percent output until the first part fails. Immediately, production falls 25 percent and remains at that level until the replacement part is ordered and arrives from the supplier. After the failed part is replaced, production once again returns to 100 percent.

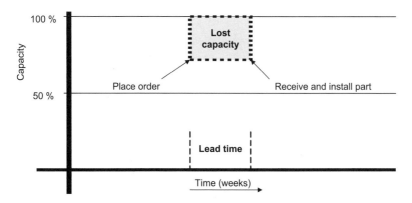

This example illustrates: • four parts in service (each carries 25% of capacity)
• no spares in storeroom
• only one failure during replenishment lead-time

Figure 2.2 Failure of a Rarely-used Item

But what happens if a second part fails before the first replenishment arrives? (Remember that four parts are in service.) Figure 2.3 depicts this situation. Depending on when during the replenishment period the second failure occurs, production will drop to 50 percent of normal capacity, then return to 75 percent when the first replenishment arrives, and finally regain 100 percent when the second replenishment arrives. In turn, what happens if a third part fails before the first replenishment arrives? You get the idea.

The decision of how many spares to stock can get very complicated. Therefore, complex algorithms based on probability theory have been developed to solve the problem. These algorithms rely on statistical distributions when trying to replicate part failure rates, then use the distributions to help calculate how many replacements to stock. Other powerful techniques first developed by the space program and military combat readiness are also used.

2.4   DISTRIBUTIONS

Distributions are nothing more than a display of all the possible values a variable can take and how often those values occur. Many different types of distributions are possible. Figure 2.4 shows two examples: a flat distribution and a bi-polar distribution of student grades in a math class. Distributions like these are unlikely to occur because seldom would the number of students getting each particular grade be the same, nor would all students get either an A or F. Although unlikely, there is nevertheless some small probability these distributions could happen.

Two more likely distributions are shown in Figure 2.5. The skewed distribution is common if the population it depicts is small. When the sample size is large, a Chi-Square distribution is fairly common. Because it can be used to help predict the probability of failure in the future using past failure rate history, the Chi-Square statistic is shown in more detail in Figure 2.6. Notice there is a different distribution for each possible value of a quantity called **degrees of freedom**.

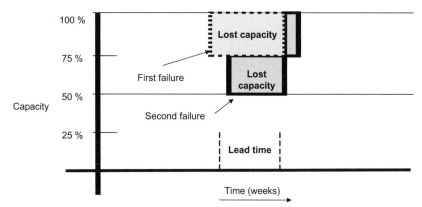

This example illustrates: • four parts in service (each carries 25% of capacity)
• no spares in storeroom
• second part fails before first replacement arrives from vendor

Figure 2.3 Failure of a Second Rarely-used Item

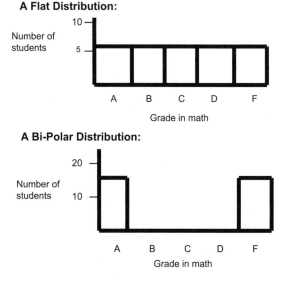

Figure 2.4 Distributions Take Many Forms

A Skewed Distribution:

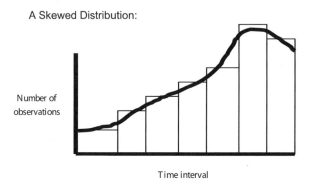

Number of
observations

Time interval

A Chi-Square Distribution:

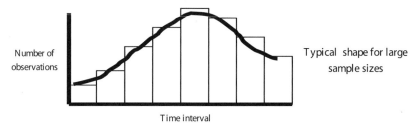

Number of
observations

Typical shape for large
sample sizes

Time interval

Figure 2.5 Other Possible Distributions

The Chi-Square distribution can be used to help predict the probability of failure
in the future using past failure rate history.

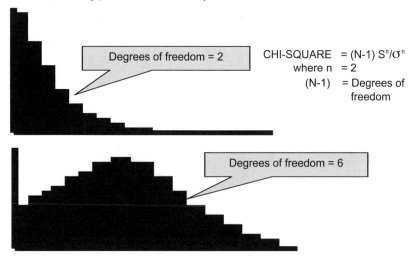

Degrees of freedom = 2

CHI-SQUARE $= (N-1) S^n / \sigma^n$
where $n = 2$
$(N-1)$ = Degrees of
freedom

Degrees of freedom = 6

Figure 2.6 The Chi-Squared Distribution

Suppose you record the number of failures, N, of a particular spare part over each of the last five years. If you determine the mean of the five observations, subtract the mean from the value of each observation, and then add the results, the sum obtained will be zero. This will be true regardless of how many observation you make. Once N-1 of the differences are known, the value of the last can be calculated. We call N-1 the degrees of freedom. Observe in Figure 2.6 that as the degrees of freedom increase, the distribution becomes more skewed to the right.

Another distribution we need to discuss is the normal distribution shown in Figure 2.7; this distribution can come into play when determining stocking levels. The normal distribution has several useful properties: 1) the shape of the curve is symmetrical, 2) the mean is directly in the middle, and 3) a specific percent of the values fall within one (68%), two (95%) or three (99.7%) standard deviations (sigma, σ) of the mean. Because it is a very common statistic, we show in Figure 2.8 how this standard deviation is calculated.

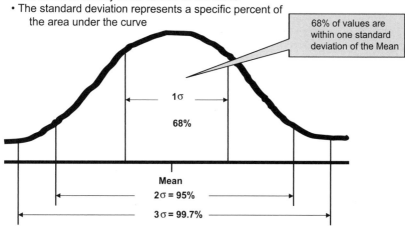

Figure 2.7 Properties of a Normal Distribution

$$S = \sqrt{\frac{\text{Sum } (X - Xavg)^2}{N - 1}} \qquad \text{(Eq 2-1)}$$

Where:

S   = standard deviation of a sample of values (also called σ)

X   = individual value in data set

Xavg = average of all values in data set = 55 in this example

N   = number of individual values in data set = 10

x	64	58	49	54	52	59	54	61	49	50
X-X avg	+9	+3	-6	-1	-3	+4	-1	+6	-6	-5
(X-X avg)^2	81	9	36	1	9	16	1	36	36	25

S = square root of 250/9 = 5.27

Figure 2.8 Calculating the Standard Distribution

When the number of parts in service is large and the probability
of failing is low, a distribution
known as the Poisson distribution can be used to estimate failure.

The Formula:

$$\textbf{Poisson} = \frac{\lambda^x \, e^{-\lambda}}{X!} \qquad \text{(Eq 2-2)}$$

where:

λ = n p
n = number of units in service
p = probability of unit failing during the year
x = number of units failing during the year

Figure 2.9 Calculating Failure Rate

The final distribution we want to discuss is the Poisson distribution, which has been found to apply better than most distributions for certain waiting-line situations, like queuing up at a storeroom counter (see Figure 2.9). The following case will better explain the use of the Poisson distribution.

 CASE STUDY 2-1 AJAX REFINING

The Situation:

AJAX Refining is a large petroleum company operating six refineries throughout the United States and Canada. In a new program to anticipate failures of key production equipment the company started calculating the probability of various parts failing. The company planned to use this information in order to upgrade its maintenance practices by having necessary spares when needed. The company statistician was called in to help with the solution.

The Proposed Solution:

The statistician proposed using the Poisson distribution to calculate the probability of failure for a commonly-stocked valve used at all the refineries.

The Numbers:

A study determined that AJAX had 2,000 identical high-pressure steam valves currently installed at the six refineries. Through a study of past failures, the probability of any one valve failing during the year was determined to be 0.002, or one chance in every 500 valves. Using standard tables of Poisson probabilities for different λ and n values, the probability that zero, 1, 2 up to 10 valves would fail during the year was determined (see Figure 2.10). When a probability histogram was plotted, the results showed a skewed distribution with a peak failure probability of 0.196 for four value failures (see Figure 2.11).

The Conclusion:

The statistician and the supply-chain manager agreed that the current stocking of 4 spare values was adequate.

An Example Using the Poisson Distribution:

AJAX Refining had 2,000 identical valves installed at six refineries. If the probability is 0.002 that any one will fail during the year, what is the probability that 0, 1, 2 up to 10 will fail during the year?

$$\lambda = n\,p = (2000)\,(0.002) = 4.0 \qquad \text{(Eq 2-3)}$$

From tables of Poisson probabilities for different λ and n values:

$\lambda = 4.0$

0	1	2	3	4	5	6	7	8	9	10
0.018	0.074	0.146	0.195	0.196	0.156	0.104	0.060	0.030	0.013	0.008

Probability of four valves failing

Figure 2.10 Calculating Failure Rate

The AJAX Refinery example shown as a probability histogram:

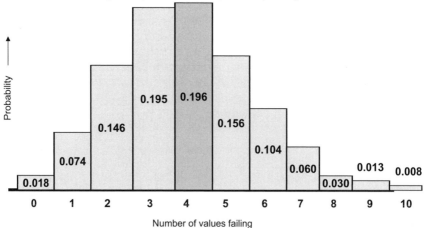

Figure 2.11 Calculating Failure Rate

2.5   PREDICTING FUTURE DEMAND

Different approaches must be used to predict future demand for spare parts depending on whether the part is an active or rarely-used item. For this discussion, we have defined the cut-off for rarely used items as one issue per month or less, on average. In the last chapter we showed that the one-per-month cut-off typically captured 80 – 90 percent of production storeroom items.

The Technique	Some Thoughts
• $\sqrt{N}$ Square root of N where N = number of parts in service	• Usually results in more over-stocking than understocking
• Seat-of-the-pants	• Usually get typical cost bias
• Decision support tools	• Get better balanced inventory plan
• Just stock one of everything	• Get major stocking errors: One correct amount about 42% Two or more correct about 55% Zero correct amount 3%

Figure 2.12 Different Approaches for Setting Stocking Levels for Slow-movers

2.5.1   Predicting Future Demand for Rarely-Used Items

Estimating future demand for slow-moving, randomly-failing parts can be a challenging exercise. Until probability theory and risk-based solutions became available, several different approaches were used to try to solve the problem (see Figure 2.12). The square-root of N technique was simple, but it tended to overstock most items; the seat-of-the-pants approach led to cost biases; the use of decision support tools gave the best balanced inventory; and the "just stock one of everything" method produced wrong stock levels about 60 percent of the time. All of these approaches attempted to answer the question, "How do I come up with a prediction for next years demand"? Figure 2.13 illustrates the dilemma. In this example, the usage over the last three

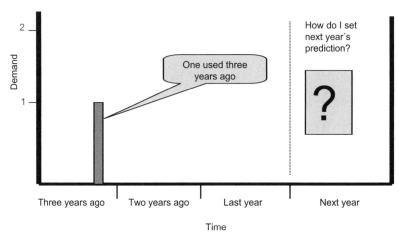

Figure 2.13 Predicting Non-forecastable Demand

The Procedure:
1. **Determine the average usage rate over the years of data.**
2. **Use the Chi-Square statistic to determine a failure rate for the future.**
3. **Set the degrees of freedom and confidence level for the Chi-Squared statistic.**
4. **Use a confidence level sufficient to anticipate an average future**
    usage rate 20-30 percent greater than the actual experienced usage rate.

Figure 2.14 Predicting Non-forecastable Demand

years was only one unit issued three years ago. Because there was no demand during the last two years, would a prediction of zero demand next year be reasonable? Maybe, maybe not.

A better approach is to use the **Chi-Square** statistic to help us. The procedure is outlined in Figure 2.14. Note that the average usage rate from step 1 can be called the **Observed Failure Rate**. As an example, it may be one-third of demand per year. The failure rate calculated in step 2 is the **Predicted Failure Rate**.

In Step 3, the degrees of freedom relates to the years of history. The confidence level indicates the probability that the future actual failure

---

The Procedure:

1. Select numerous forecasting models.
   - Moving Average
   - Linear Regression
   - Least Squares
2. Use each model with the demand data from Year -2 to forecast the know demand for Year -1.
3. Select the forecast model that has the best fit (lowest forecast error). Use this model to predict the Year -1 data.
4. Use the best fit model and the known demand from Year -2 and Year -1 to predict the demand for next year.

Figure 2.15 Predicting Forecastable Demand

rate will be less than the predicted failure rate. Using a sufficiently high confidence level will prevent the failure rate for next year being set at zero, which would otherwise invalidate the Economic-Order-Quantity (EOQ) equation discussed in Chapter 4. As the years of data increase, Chi-Square inflates the observed failure rate less. We have found that inflating the past usage data using this method produces recommended stocking levels for rarely-used items that are conservative, but not overly so.

2.5.2   Predicting Future Demand for Active Items

Active items can be much easier to work with because they can be forecasted. It's important to test forecasting models in order to find the one that does a reasonable job of predicting next year's demand. Figure 2.15 shows the typical procedure for forecasting active items:

- Data from two years ago is run though each forecasting model to see which model does the best job of predicting last year's actual demand (the model that has the lowest forecast error).
- This best fit model and the actual demand from both of the last two years are then used to predict the demand for next year. (Figure 2.16 shows this process graphically).

2.5.3   Forecasting Models for Active Items

In this section we will discuss four of the most popular models for forecasting active items.

Trend and Cyclic Models

Figure 2.17 shows data exhibiting both an upward trend and seasonality (Graph 1). The seasonality fluctuations can be removed, as shown in Graph 2.

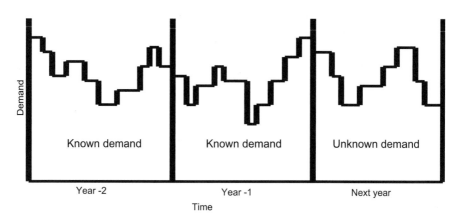

Figure 2.16 Predicting Forecastable Demand

When observations exhibit trend and seasonality influences, it is possible to adjust forecast for these factors.

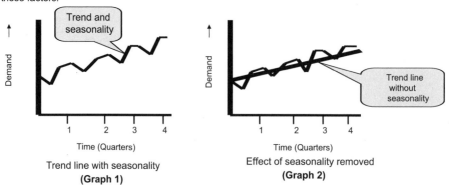

Figure 2.17 Trend and Cyclic Models

The linear regression method uses observed data and attempts to find a line of best fit, using the least squares technique. The best fit line takes the form:

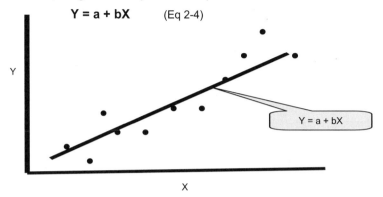

Figure 2.18 The Linear Regression Model

Linear Regression

This model uses the least squares technique to find the best fit curve that represents the data (Figure 2.18). Once the best fit is obtained, the equation can be used to calculate and apply the data.

Moving Average Models

These are a series of models, each based on the frequency used in updating the data (six-month, one-year, etc.). Figure 2.19 shows how a six-month moving average model handles predictions for future months. Data from the first six months are averaged to provide a fore-

Moving averages are updated as new information is received. The most recent observations are used to calculate an average, which is used as the forecast for the next period.

Month	Actual Demand (Units)
1	106
2	98
3	110
4	104
5	107
6	101
7	105

Using The Six-Month Moving Average:

Month 7 Predicted Demand = (106 + 98 + 110 + 104 + 107 + 101) ÷ 6
= 104.3

Month 8 Predicted Demand = (98 + 110 + 104 + 107 + 101 + 105) ÷ 6
= 104.2

Figure 2.19 The Moving Average Model

This model uses a weighted average of past data as the basis for the forecast. The greatest weight is given to the most recent data and the least weight to the oldest data. The formula is:

$$\textbf{Fnew} \; = \; \textbf{[Aold]} \; + (1 - \;)[\, \textbf{Fold}] \qquad \text{(Eq 2-5)}$$

Where:

Fnew = Exponentially smoothed average to be used for forecast

Aold = Most recent actual data

Fold = Most recent smoothed forecast

= Smoothing factor

Figure 2.20 The Exponential Smoothing Model

Month	1	2	3	4	5	6	7	8	9	10
Actual Demand	110	114	108	116	120	110	120	124	112	124

Start by getting an initial forecast. Use a six-month average and $\alpha = 0.40$

F7 = (110 + 114 + 108 + 116 + 120 + 110) ÷ 6 = 113.00

F8 = (α)(A7) + (1-α)(F7) = (0.40)(120) + (0.60)(113.00) = 115.80

Similarly:

F9 = (α)(A8) + (1-α)(F8) = 49.60 + 69.48 = 119.08

Different values of α may be used to improve the forecast and lower forecast error.

Figure 2.21 Exponential Smoothing

cast of 104.3 units for month 7. Next, the data from months 2 through 7 are averaged to predict month 8 (104.2), and so on.

Exponential Smoothing

This model is more complicated than some of the others. It differs by using a smoothing factor to assign the greatest weight to the most recent data and the least weight to the oldest data. The model works best when there is randomness and no seasonality of the data. Figure

Model 6A Air Compressor

	J	F	M	A	M	J	J	A	S	O	N	D
Actual sales (A)	101	92	96	104	99	98	102	107	100	96	95	96
Forecast sales 3-Month Moving Average (F)	103	89	97	100	101	102	106	105	103	92	98	96
Forecast sales Exponential Smoothing (F)	104	92	95	98	103	103	103	107	96	95	95	92

Calculating the MAD and MSE: Error (E) = Actual − Forecast = A - F

Parameter	3-Month moving average	Exponential smoothing
Sum of absolute values of E |E|	32	29
|E |for each month squared	104	121
MAD = |E| / N	32 / 12 = 2.66	29 / 12 = 2.42
MSE = |E| squared / (N-1)	104 / 11 = 9.45	121 / 11 = 11.00

Figure 2.22 Calculating Forecasting Errors

2.20 shows the formula used in the model, and Figure 2.21 shows a sample calculation. Different values can be used for the smoothing factor to improve the forecast and lower forecast error.

Calculating Forecasting Error

We've talked about finding the best model that results in the lowest forecasting error. Two common ways to calculate error are the mean absolute deviation (MAD) and the mean squared error (MSE). Figure 2.22 shows how each is calculated. Each method produces a different answer, but that doesn't matter as long as each method is applied consistently. Lower is better.

2.6   THE CONCEPT OF SERVICE LEVEL

With the concept of service level, we start to get into the issue of risk. Simply stated, a 95-percent service level indicates that out of every 100 requests at the storeroom counter, you can expect to get your request filled 95 times. The other 5 percent of the time, you're out of luck, and will have to do without the item until the next replenishment

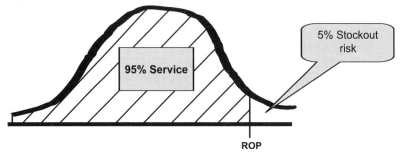

For a typical normal distribution, a specific service level is represented by
the shaded area left of the reorder point (ROP).

Figure 2.23 What Do We Mean by Service Level?

arrives. Thus, at a 95-percent service level, you are willing to accept a
5-percent risk that the storeroom will not have the part (see Figure
2.23). Common service levels are 90%, 95%, and 99%. These levels are
accompanied by stockout risks of 10%, 5%, and 1% respectively. The
level is normally set recognizing the importance of the part to support
production or some other task. It can, of course, be set at different levels for different parts, or the same level for all parts (see the case study
below).

Although the concept of service level works fine for active items, it
is an irrelevant concept for rarely-used items because of their low usage. Most of the rarely-used items in a production storeroom have seldom had one or two demands per year, let alone 100.

Bushing

CASE STUDY 2-2: AJAX ENTERPRISES

The Situation:
AJAX Enterprises provides a variety of electrical and mechanical components to the general industry. Recently the company started production of a new pressure monitoring recorder that consisted of 37 individual components, all drawn from stock at the storeroom. It didn't take long for the production manager to notice that two bearings, used individually and also as a set, were out of stock more than he thought they should be. He commented to the supply-chain manager, "I know I agreed to a 98 and a 95 service level for the left and right bearings, but we don't seem to be meeting those levels. What's wrong?"

The Proposed Solution:
The supply-chain manager recognized that each part was used at a different demand rate. He decided to calculate a Weighted Average Service Level, as shown in Figure 2.24.

The Numbers:
After looking at past records, the manager determined that the demand for the left bearing was three times the demand for the set, and 1.5 times the demand for the right bearing. When the service level was multiplied by the demand, a Weighted Service Level Index was determined.

The Conclusion:
The Weighted Average Service Level for the bearings was determined to be 96.2 percent. By raising the service level for the left bearing to 99% and the right bearing to 98%, a rate acceptable to production (98.3) was achieved.

How to calculate a weighted average service level:

This example's inventory has two non-interchangeable bearings that can also be issued as a set. Service levels must be set high for individual items to achieve an adequate service level on requests for multiple items:

Part	Current service Level	2004 demand	Weighted demand factor	Weighted service level index
Left bearing	(.98)	6	(6)(.98)	5.88
Right bearing	(.95)	4	(4)(.95)	3.80
Bearing set (one of each)	(.98)(.95)	2	(2)(.95)(.98)	1.86
Weighted average service level				11.54/12 = 96.2

Figure 2.24 Service Level for Multiple Items

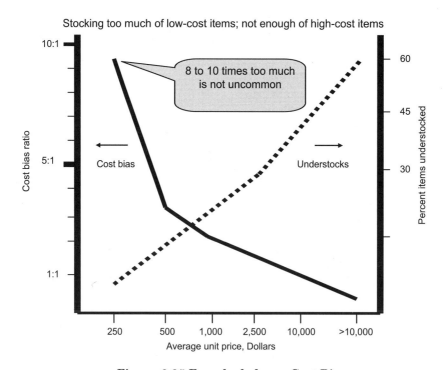

Figure 2.25 Everybody has a Cost Bias

This case study shows that if you want high service levels for combinations of parts, you must have very high service levels for the individual parts (piece parts). This relationship is also true for rarely-used items, as we will discuss in later chapters. For example, if a component has 50 individual parts, and each part has a 95-percent availability, the overall component availability will actually be less than 8 percent.

2.7   ASSESSING RISK

In 1984, Inventory Solutions Inc. (ISI) first introduced the concept of risk-based techniques to set stocking levels for non-forecastable rarely-used production spare parts. The benefits of using this technique have been dramatic, as measured by both availability improvement and net stock reductions. Now, for the first time, tools were available to calculate stocking levels for slow-moving items and replace the seat-of-the-pants techniques previously used. These older techniques led to the cost bias in stocking spares that has been observed worldwide in every storeroom, in every industry.

A cost bias is the tendency to overstock less-expensive spares and understock expensive spares (see Figure 2.25). As the tendency to overstock decreases, the likelihood of understocking increases until it is common to see spares priced at $10,000 or more at least 50 percent

---

There are two risks inherent in any stocking decision:

Getting Caught Short:

    The probable amount of time each year a part will be needed, but not in stock, times the cost implications of its unavailability.

Getting Caught Long:

    The average inventory value waiting for demand that does not occur.

    The sum of these two risks is the *Total Annual Risk Cost* for any given stock level decision.

Figure 2.26 Risk Assessment

The chart below illustrates the risks associated with an out-of-balance MRO storeroom:

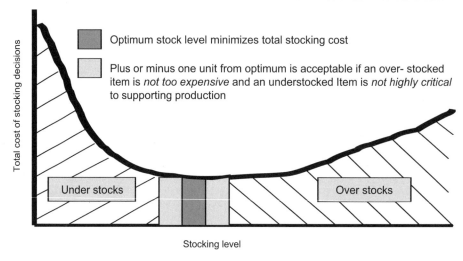

Figure 2.27 Strive for a Balanced Inventory

understocked. As a result, two risks are inherent in any stocking decision, as summarized in Figure 2.26: 1) the risk of getting caught short, and 2) the risk of getting caught long. The sum of these two risk costs is the total annual risk cost for any given stock level decision.

The end objective is to get a balanced inventory, one without too much overstocking nor too much understocking. Because the input data used to feed stocking algorithms are not always accurate, and other biases come into play when stocking spares, it is unreasonable to expect 100 percent acceptance of any result based on a mathematical computation. For this reason, we consider any inventory item that is set at a stocking level within one unit of optimum to be essentially balanced, as shown in Figure 2.27. Plus or minus one unit from optimum is acceptable if an overstocked item is not too expensive and an understocked item is not highly critical to supporting production.

One final point before we look at a case study concerning risk. Just how much conservatism should be built into stocking recommendations? We consider four factors when setting the degree of conservatism included in the MIN/MAXs recommended to clients:

Inflating the Observed Failure Rate. Use a Chi-Square value high enough to increase the observed failure rate by 20 to 30 percent.

Issues. Assume all issues (demand) on the item at the store room counter are due to the part failing in service (this assumption is a highly-conservative one).

Lead Time. Assume no expediting to improve normal delivery.

Parts in Service. Redundant, piped-in spares are considered subject to failure just like any other part. Because of its infrequent use, backup equipment may not function when called on to replace failed equipment.

By combining this conservatism to our stocking recommendations, we are probably routinely adding an average of about one-half a unit to the suggested reorder point. Even with that, however, the amount of excess inventory typically found is so large that it makes the issue of conservatism moot. Only a couple of clients over the years have worked-off or disposed of their excess inventory; afterwords, they decided to cut back further by removing some of the conservatism from the stocking levels.

150hp Motor

CASE STUDY 2-3: AJAX CHEMICAL

The Situation:
The maintenance manager and supply-chain manager at the Baton Rouge plant were discussing the stocking level for a 220-volt pump. Although they could agree that the pump was critical to operation, and some capacity might be lost if it was not in service, they had no idea what the financial impact would be. Should they increase the current reorder point, now set at 3, or not?

The Proposed Solution:
The supply-chain manager proposed that they run two cases through the risk-based software they had acquired from a consultant. The first case would assume no financial impact, and the second case $100,000 per week impact.

The Numbers:
Fig.2.28 shows the recommended stocking level for the case of no financial impact. Because the part was highly critical, the desired availability was set at 99.9 %. The recommended stocking level was a MIN. of 3 units and a MAX. of 4 units, the same as the current system. When the second case was run with the $100,000 per week backorder cost, the recommended stocking level increased to a MIN / MAX of 4/5 (Fig. 2.29). Stocking to this level produced the lowest total cost (Fig. 2.30).

The Conclusion:
The decision was made to increase the current stocking level to a minimum of 4 and a maximum of 5.

Item costs $2,000
Part is considered highly critical (Need = 99.9%)
Don't know backorder cost

Stock Level

Min	Max	Availability (%)	Avg. units	Inventory value ($)	Total cost ($) (At 25% carrying)
-1	0	57.9	0.00	0.00	0.00
0	1	89.2	0.57	1,137	284
1	2	98.0	1.46	2,910	728
2	3	99.7	2.43	4,869	1,217
3	4	99.9	3.43	6,863	1,716
4	5	99.9	4.43	8,863	2,216
5	6	99.9	5.43	10,863	2,716

(Optimum stock level — pointing to 3/4 row)

Figure 2.28 Risk Assessment

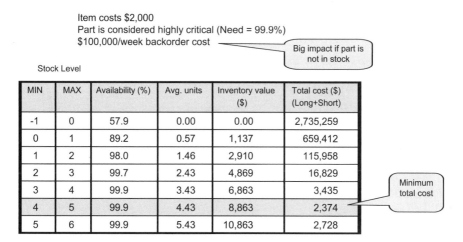

Item costs $2,000
Part is considered highly critical (Need = 99.9%)
$100,000/week backorder cost ⟶ Big impact if part is not in stock

Stock Level

MIN	MAX	Availability (%)	Avg. units	Inventory value ($)	Total cost ($) (Long+Short)
-1	0	57.9	0.00	0.00	2,735,259
0	1	89.2	0.57	1,137	659,412
1	2	98.0	1.46	2,910	115,958
2	3	99.7	2.43	4,869	16,829
3	4	99.9	3.43	6,863	3,435
4	5	99.9	4.43	8,863	2,374
5	6	99.9	5.43	10,863	2,728

Minimum total cost

Figure 2.29 Risk Assessment

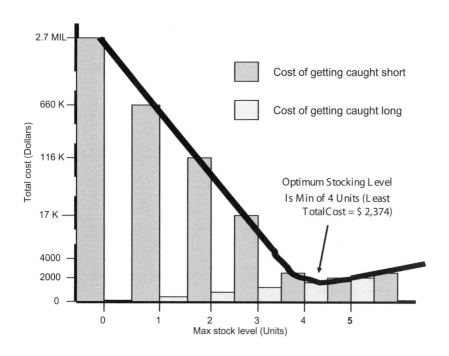

Figure 2.30 Risk Assessment

Let's look at the implications of this case in more detail. Figure 2.29 shows the cost impact of taking too much risk stocking the critical pump. A decision to not stock the item (MAX = 0) and only order after a demand has occurred results in an availability of only 57.9 percent, all coming from the probability of not having a failure over the next lead time interval. The annual inventory risk cost for this decision is very easy to calculate as there will never be any inventory of this item. However, the annual backorder risk cost is very high ($2,735,259). With a zero stock decision, the duration of a backorder is always equal to the lead time.

For stock levels greater than zero, backorder durations are less than a full lead time because one or more units have been ordered prior to the backorder demand. As the stock level is increased, the inventory risk cost increases and the backorder risk cost decreases until a minimum cost is reached at a MIN of 4 and a MAX of 5. Above a MAX of 5, the increasing costs of carrying the resulting on-hand and not-needed inventory more than offsets the resulting reductions in backorder risks. The best decision is the stock level which has the lowest total annual risk cost. Risk-based assessment techniques are not only highly effective in managing existing spare parts, but are also an equally effective tool for setting stocking levels of initial spares, as will be shown in Chapter 6.

2.8   THE MERITS OF SENSITIVITY ANALYSIS

The data needed to calculate stocking levels is frequently poor or otherwise unreliable for many reasons:

- "We haven't bought it for several years and our normal supplier is no longer in business. I don't have a clue what the current lead time should be."
- "They ask what risk we should accept of running out of this part so they can set the criticality for the item. Nobody around here seems willing to make that call."

- "We need an estimated failure rate to set the MIN/MAX for the item. Maintenance is tied up on an emergency, so they can't help."

When data are unreliable and you need an answer, the solution is to run a sensitivity analysis on the item. Use a range of one or more input variables. In the remainder of this section, we will show some typical sensitivity analyses.

Part cost $85   Carrying cost 20%/yr
Desired availability 99.9%   Order cost $75.00

Lead time range (weeks)	MIN needed	MAX needed	EOQ
0 - 0	-1	0	3
1 - 1	0	3	3
2 - 7	1	4	3
8 - 17	2	5	3
18 - 30	3	6	3
31 - 46	4	7	3
47 - 63	5	8	3
64 - 82	6	9	3
83 -103	7	10	3

Figure 2.31 Sensitivity to Lead Time

Part cost $200   Carrying cost 20%/yr   P.O. cost $75.00

Availability	Days at risk	MIN needed	MAX needed	EOQ
87	47	-1	0	1
99	4	0	1	1
99.95	0.2 (5 hrs)	1	2	1
99.99	0.04 (1 hr)	2	3	1
99.995	0.02 (30 min)	3	4	1

Availability is the combination of first having a failure and then having sufficient stock in the storeroom to meet the demand.

Figure 2.32 Sensitivity to Criticality

2.8.1 Sensitivity to Lead Time

The replenishment lead time is a value usually determined by the combination of internal paper processing time, vendor shipment response time, and delivery and receipt time. Any or all are subject to change and usually do. Smart suppliers always quote a shipment lead time that is conservative and can almost always be met. Furthermore, a strike at the supplier plant could invalidate even the latest delivery quote.

Figure 2.31 shows how the MIN and MAX for an item can vary as the lead time to replenish increases. The value of this analysis is the ability to show the wide range over which the lead time can vary and still stock the same amount of inventory. In Chapter 3 we will discuss lead time further, and introduce a concept known as the lead time bias.

2.8.2 Sensitivity to Criticality

The risk of running out of a spare is directly related to the criticality assigned to the part. We suggest setting criticality at only three levels: high (99.9 % availability), medium (99.0 %), and low (97.0 %), but any other level could be assigned. Figure 2.32 shows the days at risk and the minimum stock level required for various availability levels. Remember, availability is the combination of first having a failure, and then having sufficient stock in the storeroom to meet the demand. Probability theory comes into play in determining the chance of having a failure, after which the availability level chosen determines whether or not the item will be in the storeroom when requested.

2.8.3 Sensitivity to Mean Time Before Failure

Because initial spares (new spares) lack usage history unless they can be supplied from some other source, estimates of Mean-Time-Before-Failure (MTBF) must act as a surrogate for past usage when setting stocking levels. Figure 2.33 shows how MTBF levels can influence stocking levels. Maintenance personnel are usually the best source for these estimates unless the supplier of the spare has data and is willing to provide it.

2.8.4   Multiple Sensitivity Analyses

It is often useful to look at the sensitivity of MIN/MAX levels relative to more than one input parameter at a time. Figure 2.34 shows the
sensitivity of the MIN/MAX to both criticality and lead time. In this
example, the current 1/2 MIN/MAX is valid for only six of the fifteen
possible combinations. For seven combinations, the current levels
could be decreased and for two they must be increased.

Part cost $200   Carrying cost 20%/yr
Desired availability 99.9%   Order cost $75.00

Mean time before failure (MTBF) (years)	MIN needed	MAX needed	EOQ
809 - 17.4	0	1	1
17.3 - 8.4	1	2	1
8.3 - 4.1	1	3	2
4.0 - 3.1	2	4	2
3.0 - 1.9	2	5	3
1.8 - 1.6	3	6	3
1.5 - 1.1	3	7	4
1.0 - 1.0	4	8	4
0.9 - 0.7	4	9	5

Figure 2.33 Sensitivity to Mean Time Before Failure

Part cost $850   Order cost $75.00   Carrying cost 20%

Current system LT = X weeks	Current MIN = 1	Current MAX = 2

Criticality	Lead Time				
	X - 9	X - 6	X	X + 10	X + 20
(Availability)	MIN/MAX	MIN/MAX	MIN/MAX	MIN/MAX	MIN/MAX
High (99.9)	0/1	1/2	1/2	2/3	2/3
Medium (99.0)	0/1	0/1	0/1	1/2	1/2
Low (97.0)	0/1	0/1	0/1	1/2	1/2

☐ MIN/MAX could be reduced   ▨ MIN/MAX okay   ▨ MIN/MAX needs to be raised

Figure 2.34 Sensitivity to Lead Time and Availability

Evaluate the sensitivity of stocking a part to availability, lead time, and demand simultaneously

Minimum availability = 99.0%

Total usage	Lead time (Weeks)			
	X-8	X-5	X	2X
0	0	0	0	0
1	0	0	0	1
2	0	0	1	1
4	0	1	1	2
8	1	2	2	4
16	2	3	4	6
24	2	4	5	9
32	3	4	6	11
40	3	5	7	13
48	3	6	8	15

Figure 2.35 Sensitivity to Multiple Parameters

Figure 2.35 shows a similar analysis when the availability is set at medium (99.0 %) and the MIN is determined against both lead time and annual demand.

Even more complex multiple-sensitivity analyses are possible. For example, when making initial spares buy decisions, it is possible to have lookup tables prepared in which the Number-in-Service, MTBF, and Lead Time can all be varied over a range for a specific level of Criticality. These tables help plant management to set the buy amount once they determine the criticality for the part, the expected number in service (usually available from the bill of materials), the estimated MTBF (usually available from the vendor or estimated by maintenance), and the projected lead time (a vendor input).

The advent of risk-based assessment allows all of these analyses to be completed quickly, and provides plant management with information that supports making better stocking decisions. When better decisions are made, risks to production are reduced and the money invested in stocking inventory is optimized.

CHAPTER 3

SETTING THE REORDER POINT

3.1 WHAT THE READER SHOULD LEARN FROM THIS CHAPTER

- The key parameters for setting the reorder point (ROP)
- How different parameters affect the ROP
- What is a lead time bias?
- How to set the ROP for an active item
- How to set the ROP for an active item with cyclic demand
- How to set the ROP for slow-moving items
- How the incremental max will tell you when "enough is enough"

3.2 THE CONCEPT OF THE REORDER POINT

The concept of using an order point, or reorder point, to trigger the procurement of an item is not unique to production or manufacturing processes. If you think about it, we use reorder points every day in our personal lives. For example, we reorder our car license plates before they expire. Likewise, we reorder checks from the bank before we use the last one. In most cases, failure to order on time is not likely to cause a major financial crisis, but at worst a minor inconvenience.

Not true in industry. Failure to replenish raw materials or spare parts on time can cause production losses, violate regulatory or environmental laws, and compromise the safety of workers. This is especially true for spare parts that are determined to be critical for sustaining production or other important plant activities. The usual recourse to offset late ordering is to have the purchasing department expedite delivery, often at substantial additional cost. One such incident happened at an eastern U.S. nuclear plant that lacked an $800 spare part needed to bring the plant back into operation after a shutdown. To expedite delivery and avoid further loss of power production, a private jet was sent to retrieve the part from the west coast supplies at a cost of nearly $20,000!

The important thing to remember about the reorder point (ROP) is that it is the point in the replenishment process that should trigger the normal ordering of a replacement stock keeping unit (SKU). Setting the reorder point at the proper level is what protects the availability of equipment to sustain the production process. Some material management systems set a reorder point, but then recommend actually ordering the SKU one level below the system ROP. The bottom line is that the reorder point is the point at which you trigger the reorder, regardless of where you set it or what you call it.

3.3   WHEN TO USE REORDER POINT/ REORDER QUANTITY

Most items can best be replenished by using a reorder point/reorder quantity approach when ordering. Most applications that fit this criteria are for items that can be procured independent of other items. Examples are:

- raw materials that are purchased in standard quantities such as, by the drum, the case, or 10-foot length; frequently these items may be bought in the standard quantity and then distributed locally in smaller amounts (e.g., bought by the drum, but distributed from stores by the gallon).

- production finished goods such as sub-assemblies.

- maintenance spare parts.

- retail and wholesale purchasing decisions (purchased by a unit of measure called "each).

Other items do not fit the use of ROP/ROQ:

•   bulk materials used in a fabrication process

•   small pieces and parts used in an assembly process (nuts and bolts).

3.4   KEY STOCKING INPUT PARAMETERS

Most computer codes for setting stocking levels use algorithms that depend on specific input parameters to establish the recommended order point for items, whether active or rarely-used. Figure 3.1 shows the key input parameters needed to set the reorder point (also known as the Minimum, MIN) and economic order quantity (EOQ) for a rarely-used SKU. A detailed discussion of the key parameters follows. Referring to Figure 3.1, note that the MIN is influenced only by the part criticality, lead time, set size, and annual demand. Of these four parameters, only demand comes into play when determining the EOQ (see Chapter 4 for more on setting the EOQ).

Important Inputs:	Typical Outputs:
• *Criticality*	• *Minimum stocking level*
• *Lead time to replenish*	• *Economic order quantity (EOQ)*
• *Issued in sets of*	• *Maximum stocking level*
• *Usage (Demand)*	

How to interpret the recommended stocking levels:

MIN – the minimum recommended stocking level based on criticality / lead-time / sets of / demand

EOQ – the text book order quantity based on annual usage/ purchase order cost / carrying cost / price

MAX – the maximum recommended stocking level based on minimum stocking level and calculated EOQ

Key point: carry no less than the minimum stock level and no more than the maximum level

Figure 3.1 Key Stocking Input Parameters

3.4.1 Criticality

Any discussion of criticality must first start with an understanding of the meaning of Availability, which is defined as the combination of first needing a spare part (because a part has failed) and second the likelihood of having one in the storeroom to meet the demand. For example, a 99% availability level means that, 99 days out of 100, either a spare part is not needed because the equipment is functioning properly or, if the spare part is required, one is available in the storeroom. Conversely, 1 day out of 100 there will be a need for the part **and** it will **not** be available. Clearly, availability is synonymous with **risk.**

Because spare part availabilities are set at the piece-part level, they need to be set extremely high to protect the overall availability of their corresponding equipment.

Table 3.1 How Piece-Part Availability Affects Component Availability

Piece-Part Availability	10-Part Component Overall Availability	50-Part Component Overall Availability
99.9%	99%	95%
99.5%	95%	78%
99.0%	90%	61%
98.0%	82%	36%

Consider Table 3.1. If all individual piece-part availabilities are set at 99.9%, a 10-part component would have only a 99% overall availability; for a 50-part component, availability would decrease to 95%. At 99.0% piece-part availability, the values drop to 90% and 61% respectively. Although a 99% service level may be entirely acceptable for an active part (see Chapter 2), a 99% piece-part availability would not be appropriate if the part is installed on an essential component (too much risk of lost production).

Not all spares are highly critical to the production process. Yes, some are so important that they must be available at all times or production will be lost. Others may be only a minor nuisance if not in stock, and replacement deliveries of several weeks can be tolerated.

Assigning only three distinct levels to part criticality is usually enough, although some companies prefer more (5-to-10 levels are not uncommon). Inventory Solutions, Inc. sets three levels, as defined in Table 3.2.

Table 3.2 Criticality Levels Defined by Inventory Solutions, Inc.

Criticality	Minimum Availability	Definition
High	99.91% (8 Hour/year risk of running out of part)	Unsafe condition or output will occur. Can not substitute another part or jury-rig around the problem. No alternative. Must have the part.
Medium	99.0% (3.65 days/year risk)	Inconvenient but can maintain production until a replacement arrives.
Low	97.0% (11.0 days/year risk)	Can substitute or go without for several weeks

Maintenance is usually responsible for setting the criticality of a spare part. Although other plant personnel may have an opinion, only maintenance personnel really know the consequences of not having the part when needed. They also are more likely to know if some other part can be substituted for the failed item, possibly lowering the criticality from high to medium, or even low. It must be remembered that assigning a criticality is a judgment call by maintenance. What's high to one person may be only medium to another. Consequently, it is not uncommon to see the percent of spares assigned to high criticality (or some other level) vary widely from plant-to-plant, company-to-company, and industry-to-industry. A typical breakdown based on millions of spares reviewed is shown in Table 3.3

Table 3.3 Typical Breakdown of Millions of Spared Reviewed

Criticality Assigned	Percent of Parts Reviewed
High	50 - 65 %
Medium	20 - 30%
Low	15 - 25%

It is not practical to assign 100% availability to a spare part because, in theory, an infinite amount of inventory would be required to protect against creating a backorder. What is clear is that as availability increases, more inventory is required to support the lower risk of

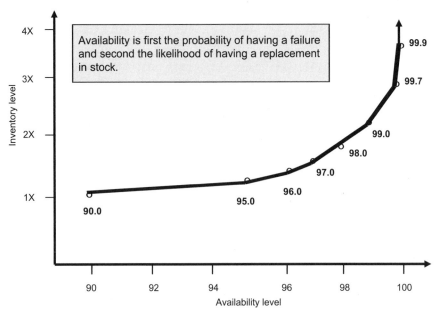

Figure 3.2 The Cost Of Improving Availability

running out of stock. Figure 3.2 graphically displays the diminishing return on investment found on the availability-inventory curve. For example, raising availability from 90.0% to 98.0% can nearly double the amount of inventory required, all other stocking parameters remaining the same. To go to 99.9% can double the required stocking again. Because the vast majority of the spare part criticalities assigned in industry are either 99.0 or 99.9, we are "playing the game" in the very steepest portion of the curve. As one knowledgeable plant manager once remarked, "as you approach 99.9% availability, the inventory required becomes asymptotic to the moon."

3.4.2 Lead Time

Experience at hundreds of storerooms shows that investment in rarely-used spare parts inventory can routinely be cut in half if suppliers could deliver the spare in two weeks or less compared to current longer lead times.

Keep in mind that lead time takes many forms. Figure 3.3 shows a typical lead-time chain from suppliers, through manufacture, warehousing, and distribution before reaching retailers and the ultimate customer. Normally, the lead-time chain starts with a requisition to

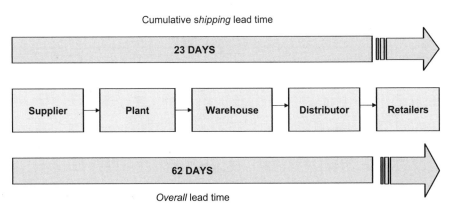

The majority of the *shipping* lead time is moving finished goods to the distributors. The majority of the *overall* lead time is the days to manufacture.

Figure 3.3 Lead Times Takes Many Forms

replenish a spare part that has hit the reorder point in the storeroom. Once the paper (yes, some of us still push paper) reaches purchasing, the requisition turns into a purchase order (or blanket order). This step may take several days, or even weeks, waiting in queue to have delivery date, price, or other information verified before issuing the purchase order. Once the spare part is received back at the plant, there may be other delays getting the part into the storeroom bin; some of these delays may be caused by receipt administration and quality inspection. Consequently, these front-end and back-end administrative/inspection delays can often equal or exceed the actual supplier's lead time to manufacture or supply many parts. Other factors, including shipping, can take up a large amount of the overall lead time as shown in Figure 3.3.

To understand better the influence of lead time on the stocking of inventory, it is useful to look at how inventory is impacted by changes in lead time. Table 3.4 highlights two significant points. First, as replenishment lead time increases, the Minimum and Maximum stocking levels also increase. Second, as the lead time gets longer, the spread in weeks over which the same MIN/Max applies widens.

Table 3.4 How Replenishment Lead Time Affects the Stocking Level

Replenishment Lead Time (Weeks)	Stocking Level	
	Minimum	Maximum
0	-1	0
1	0	1
2 – 4	1	2
5 – 9	2	3
10 – 16	3	4
17 – 25	4	5
26 – 34	5	6
35 – 44	6	7

For example, any lead time between 10 and 16 weeks (a 6-week spread) requires the same stocking level (MIN 3 and Max 4). A stocking MIN of -1 means the spare does not have to be stocked, but only ordered when there is a demand at the storeroom counter. At a 20-week lead time, the recommended stocking is a MIN/MAX of 4/5. The analysis also reveals that the lead time would have to increase to 26 weeks before more stock was required, whereas a decrease to 16 weeks could reduce required stock by one unit. A decrease of four units (MAX goes from 5 to 1) is possible if the supplier could routinely deliver the replacement spare within one week.

In Chapter 2 we saw that stocking rarely-used inventory in most storerooms results in a cost bias. Likewise, stock levels determined without using a good decision support tool also exhibit a lead time bias. Figure 3.4 shows a typical lead time bias for a production plant storeroom. For items with a short lead time, 0-2 weeks, actual inventory was overstocked by a factor of 5.6 compared to the recommended level. At a lead time of 16-20 weeks, actual and recommended stock levels were in balance (A/R=1.0). After 21 weeks there was a consistent tendency to understock versus the recommended level. Overall, the actual and recommended inventory for the 2,577 items was in balance as seen by the totals (A/R=1.0).

Although less severe than a *Cost Bias,* a lead time bias also exists at most storerooms as shown below:

Lead time (Weeks)	Number of items	Actual inventory ($) (A)	Recommended inventory ($) (R)	A/R
0 - 2	2	1,476	260	5.6
3 - 4	61	185,938	58,048	3.2
5 - 6	259	416,749	305,268	1.4
7 - 8	97	256,760	197,507	1.3
9 - 10	93	250,184	202,276	1.2
11 - 15	1,519	4,757,338	4,333,100	1.1
16 - 20	213	697,317	699,178	1.0
21 - 30	210	1,209,580	1,511,970	0.9
31 - 40	65	547,900	602,830	0.8
41 - 50	58	433,745	683,920	0.6
Totals	2,577	8,756,987	8,594,357	1.0

Figure 3.4 Everybody Has a Lead Time Bias

Simply stated, a lead time bias means that readily available parts tend to be heavily overstocked. As a result, parts that can be procured quickly are almost never needed in a hurry. Conversely, long lead-time parts tend to be understocked. Hence, Murphy's Law – the part that is back ordered is always the most difficult to expedite because it probably has the longest lead time! Table 3.5 confirms this conclusion for hundreds of storerooms. When the lead time was short (less than 6 weeks), the reorder point was set too low only 10 percent of the time. With lead times of 52 weeks or more, understocking of the reorder point occurred 66 percent of the time.

Table 3.5 As Lead Time Increases, Understocking Also Increases

Lead Time (Weeks)	Percent of Reorder Points Set Too Low
0 – 6	10%
7 – 13	24%
14 – 26	36%
27 – 52	57%
> 52	66%

Figure 3.5 shows the sensitivity of the reorder point to changes in the lead time for spare parts. The impact of increasing lead time on the MIN/ROP, MAX and average inventory are all shown. Notice that the economic order quantity (EOQ) is not impacted by the lead time changes because the EOQ formula does not contain lead time as a variable. Figure 3.6 displays the results in graphical form. Notice the stair step shape of the chart as lead time increases. As indicated, a lead time of 16 weeks requires a MIN of 5 units, a MAX of 7 units (because the EOQ is 2), and $4,503 of average inventory. Case 3-1 looks further at lead time.

Price of part = $1,000   Carrying cost = 20%
Cost of purchase order = $70.00

Lead time (wks)	MIN/ROP	MAX	EOQ	Avg. invy ($)
1	0	2	2	1,375
2	1	3	2	2,250
4	2	4	2	3,001
8	3	5	2	3,501
12	4	6	2	4,202
16	5	7	2	4,503
20	6	8	2	5,003
26	7	9	2	5,254
32	8	10	2	5,505
40	9	11	2	5,806
50	11	13	2	6,258

Figure 3.5 ROP Sensitivity to Lead Time

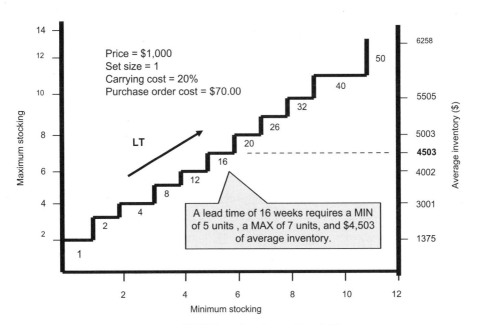

Figure 3.6 ROP Sensitivity to Lead Time

CASE STUDY 3-1: AJAX MANUFACTURING

The Situation:

AJAX Manufacturing is an assembly shop located in Charleston, West Virginia. They mainly buy parts and components from other suppliers, then assemble them into custom-made assemblies for general industry clients. The plant management is constantly on the lookout for ways to cut costs, and inventory is not exempt from scrutiny. At last week's staff meeting, the supply-chain manager raised the question, "How much inventory could we save from this list of nine spare parts if our current lead time could be cut by 25 percent?"

The Proposed Solution:

The company had access to an inventory decision support tool from its work with a consultant. The supply-chain manager, who decided to run a sensitivity analysis for the nine items, compared the current lead time against the shorter lead time that resulted from a projected 25 percent decrease in the vendor shipping date.

The Numbers:

The projected savings was $31,852 (see Figure 3.7). For example, the 250HP AC motor could have its MIN cut from 1 unit to zero if the 8-week lead time were reduced by 25 percent to 6 weeks. Also of interest is Part 037899, where the 25 percent lead time reduction suggested the part not be stocked (a -1 MIN).

The Conclusion:

Discussions between AJAX and the suppliers resulted in a 25 percent shorter lead time for seven of the nine items without affecting pricing.

Vendor: AJAX Manufacturing*

Part No.	Description	Part cost ($)	Lead time (Wks)	Current MIN	Potential MIN	Savings ($)
830031	Motor AC 250HP	19,225.00	8	1	0	19,225
845345	Main Coil	3,000.00	12	1	0	3,000
846790	Trip Card	2,681.97	10	1	0	2,682
037899	Amplifier X4T	1,160.10	7	0	-1	1,610
569873	Rectifier	1,332.73	8	2	1	1,333
337678	Brush Holder	1,261.00	8	2	1	1,261
435645	Circuit Board	1,075.23	15	2	1	1,075
344545	Impeller	913.00	8	1	0	913
768577	Voltage Regulator	752.58	7	2	1	753

*This chart assumes lead times for all parts are reduced by 25%.

Total potential inventory savings $31,852

Figure 3.7 Calculate Lead Time Reductions

3.4.3 Set Size

Recall that set size is one of the main parameters affecting inventory stocking (see Figure 3.1). Set size also happens to be one of the least understood and ignored parameters when stocking spares. Why?

Most spare parts carry a stocking designation of each, meaning they are normally purchased, priced, and stocked as a single unit of each. However, although designated as each, they are frequently bought in cases of 12 or some other number.

Consider the example of two bearings on the shaft of a pump. In the storeroom, the bearings are probably stocked as each, but when maintenance repairs the pump due to a bearing failure, they are likely to draw out two and replace both bearings of the pump, even though only one has failed. The bearing becomes a set of two. Therefore, at least two, or a multiple of two, should be stocked. What happens if maintenance draws out two bearings (a set), but only replaces one in the pump, returning the unused bearing to stores? Now we have a problem, because we are stocking only one-half of a set!

Price of part = $1,000   Availability = 99.00   Number in service = 2
Carrying cost = 20%   Cost of purchase order = $70.00

Set size	Normal MIN	Adjusted MIN	MAX	EOQ	Avg. Invy ($) for adj. MIN
1	6	6	9	3	6,055
2	8	9	10	2	8,004
4	8	11	12	4	10,503
6	12	17	18	6	15,903
12	24	35	36	12	33,724

Sometimes the MIN is adjusted to be one unit below the MAX to protect against not having a full set available if part of a set is returned to the storeroom and not used during maintenance.

Figure 3.8 ROP Sensitivity to Set Size

This practice happens frequently, causing delays in getting equipment back into service, as well as costly expedites by purchasing. Special care is needed to determine the most likely set size if a spare part is used in sets. The experience of maintenance now comes into play because they must determine what needs to be stocked as a set, and how many constitute the set size. In many cases, that decision becomes a judgment call.

Figure 3.8 shows a sensitivity analysis for a spare over a range of set sizes. Given the parameters shown (two in service, availability = 99.00), if the spare is not a set (only one unit required for routine maintenance) the MIN would be 6, the MAX 9, and the EOQ 3, resulting in an average inventory of $6,055. As a set of 2, the values become 8, 10, and 2 respectively for the MIN, MAX and EOQ; at a set size of 12, we get 24, 36 and 12.

Notice the column labeled "Adjusted MIN". By setting the MIN one unit below the MAX, we can compensate for the possibility of stocking a partial set. For example, if the set size was 2, the normal MIN would be 8 and the MAX 10 (EOQ=2). By setting the MIN at 9 units instead of 8, we have programmed the reorder point to trigger a requisition in the event maintenance uses only one unit of a two-unit set when making the repair. Yes, it is true that ordering 2 units at a MIN of 9 would cause the stock to rise to 11 units, exceeding the normal MAX of 10. However, that option is generally considered a lesser evil than not having full sets in the storeroom when needed.

3.4.4 Demand

As with criticality, not all demand (usage) is the same. Some spares (active) have high demand and are forecastable, whereas others (rarely-used) have little or no usage and are non-forecastable. We saw in Chapter 1 that it makes sense to sort usage into active and rarely-used categories, depending on the usage rate.

The usage rate of a spare part has been found to be the most important factor in determining which inventory management technique should be applied. A frequently chosen cutoff defining a rarely-used item is any item that is issued once per month or less on average. There are two reasons for setting the cutoff at this level:

- Even with the most sophisticated multi-model forecasting systems, some models fail to work properly below this rate of usage.

- Most inventory items are issued either significantly more than once per month, or are hardly used at all.

We know from years of experience in the process industries that active spare parts comprise only about 10 –15 percent of the items in inventory, with rarely-used items making up the balance. Typically, 40-60 percent of the rarely-used items at a site have had no usage in the past three years. Traditional inventory planning systems do not work for these items.

A final step in classifying spare part inventories is to sort the rarely-used items into key and non-key groups. This step is important because it determines the number of items that justify intensive management (key items) as opposed to those that can be managed with less

Stocks can be depleted at different usage rates:

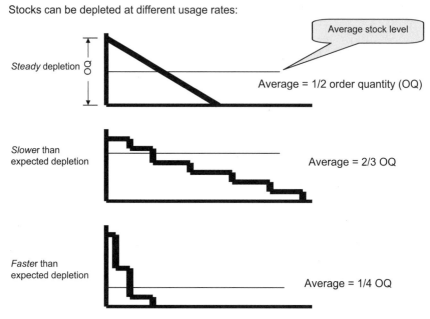

Figure 3.9 The Effect of Different Usage Rates

stringent controls (non-key items). It also shows the Pareto principle in action, where focusing on a small percentage of inventory items can cover the majority of the inventory value.

The demand for a spare part is a function of how soon it is used in the production process either due to failure, preventive maintenance, or scheduled overhauls. Also, the rate of usage is not always steady, as shown in Figure 3.9. With steady usage we expect the average stocking level to be equal to one-half the order quantity. Slower or faster usage will have a different impact on the average inventory. Figure 3.10 shows another typical situation for a rarely-used spare part. In this example, the ROP is set at 2 units and the EOQ is 2, setting the MAX at 4 units. One unit is over MAX, and once that unit is used it may be a considerable amount of time before the ROP is reached and a replenishment order is initiated. There may or may not be usage of the part during the lead time period.

Rarely-used items change stocking level slowly over time

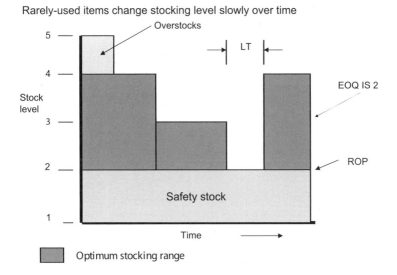

Figure 3.10 ROP for Rarely-used Items

Price of part = $1,000   Carrying cost = 20 %
Cost of purchase order = $70.00

Usage over 3 Yrs	MIN/ROP	MAX	EOQ	Avg. invy ($)
18	3	5	2	3,501
16	3	5	2	3,606
15	2	4	2	3,120
14	2	4	2	2,712
12	2	4	2	2,850
10	2	4	2	2,950
9	2	4	2	3,090
8	2	3	1	2,551
7	1	2	1	1,532
6	1	2	1	1,656
4	1	2	1	1,765
3	1	2	1	1,834
2	0	1	1	874
0	0	1	1	966

Figure 3.11 ROP Sensitivity to Part Demand

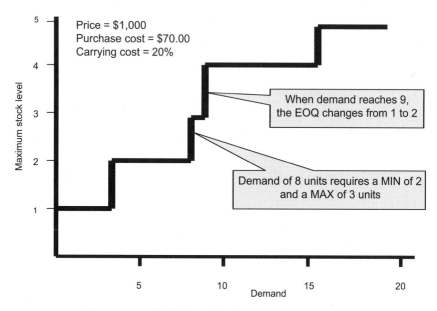

Figure 3.12 ROP Sensitivity to Usage (Demand)

The sensitivity of the reorder point to demand can be seen in Figure 3.11. In this case study, usage over a three-year period was varied from zero (no demand) to 18 issues. Reading from the bottom up, the change in MIN ranged from zero to 3 units. Once demand reached 9 units (an average of 3 per year), the EOQ increased from 1 to 2 units. Average inventory is also shown based on a part cost of $1,000. The same data are shown graphically in Figure 3.12. Notice how the change in the EOQ from 1 to 2 creates a short step in the shape of the trend line when demand hits 9 units over the 3-year period.

3.5 SETTING THE REORDER POINT

Now that that we have seen how the reorder point is impacted by the key input parameters of criticality, lead time, set size, and demand, it is time to look at the procedure for actually setting the ROP.

3.5.1 Active Items

In Chapter 1, we discussed the characteristics of active items including their tendency to have high usage, short lead times, and a need for high service levels. For these items, the approach for setting the reorder point is as follows:

- Collect several years of usage history (usually two years is sufficient).

- Select an active-item forecasting program that has many forecasting models (such as linear regression, moving average, and exponential smoothing) to predict future demand.

- Run the SKUs through each forecasting model and select for each SKU the model which does the best job of predicting the most recent years' actual demand pattern using the older years' demand history.

- Next, use this best model for each SKU, along with all years of actual demand history, to predict next year's future demand.

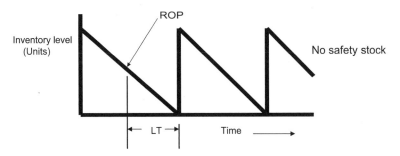

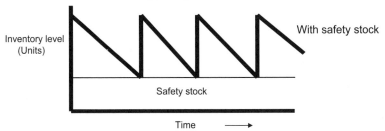

Figure 3.13 Active Reorder Point (ROP)

- Finally, run the predicted demand along with the other stocking parameters such as lead time, price, carry cost, etc. through the stocking program to set the MIN and MAX for each SKU.

In an ideal world where demand and lead time are exactly predictable, there is no need to carry safety stock when making the stocking decision. That is seldom the case, however, and so safety stock must be factored into the stocking decision, as shown in Figure 3.13. When safety stock is not included, it is necessary to set the lead time sufficiently far in advance to insure that replenishment occurs before there is a stock out. With safety stock included, the risk of stock out is typically avoided, unless there is some serious disruption in the supply chain.

Because of the higher demand for active items, it is not uncommon to have large variability in lead times. Figure 3.14 shows how the reorder point changes, for different service levels, as lead time varies. In the example, a normal distribution curve of lead times yields an average lead time of 6 weeks and a standard deviation of the lead time of 2 days (see Chapter 2 for more on distributions). Thus, at a constant demand rate of 50 units per day, and using the appropriate Z factor for a normal distribution, we get an ROP of 464 units (300 units for usage during the lead time and 164 units for safety stock) for a 95 percent service level. For 99% and 99.9% service levels, the ROP increases to 533 and 609 units, respectively, because of the need for more safety stock.

The flip side of the above situation occurs when the lead time is constant and the demand varies. In this example (Figure 3.15), the average usage rate for the year is 50 units with a standard distribution of the usage rate of 5 units. Using similar Z-factors and service levels, the reorder points range from 474 units to 496 units as service level increases from 95.0 to 99.9%. Safety stock is not an important factor in this case. Let's look at a case study involving variable, but cyclic demand.

ROP = Expected usage during lead time + Safety stock

$$ROP = (U)(LTavg) + (Z)(U)(S) \qquad (Eq\ 3\text{-}1)$$

Where: Z = Factor from normal distribution service level
S = Standard deviation of lead time = 2 days
U = Constant usage rate = 50 Units per day
$LTavg$ = Replenishment average lead time = 6 days

Parameter	95% Service	99% Service	99.9% Service
Z	1.64	2.33	3.09
Safety stock [(z)(U)(S)]	164 Units	233 Units	309 Units
LT usage [(U)(LTavg)]	300 Units	300 Units	300 Units
ROP	464	533	609

Figure 3.14 Calculating the ROP with Variable Lead Time

ROP = Expected usage during lead time + Safety stock

$$ROP = (Uavg)(LT) + (Z)(\sqrt{LT})(S) \qquad (Eq\ 3\text{-}2)$$

Where: Z = Factor from normal distribution service level
S = Standard deviation of usage rate = 5 units
$Uavg$ = Average usage rate during the year = 50 units
LT = Replenishment lead time = 9 days

Parameter	95% Service	99% Service	99.9% Service
Z	1.64	2.33	3.09
Safety stock [(z)($\sqrt{LT}$)(S)]	24 Units	35 Units	46 Units
LT usage [(Uavg)(LT)]	450 Units	450 Units	450 Units
ROP	474	485	496

Figure 3.15 Calculating the ROP With Variable Usage

CASE STUDY 3-2 AJAX REFINING

The Situation:

AJAX Refining is a small refinery in the southern United States producing mainly light-end oil for lubricants. Because the refinery runs on a 24-hour schedule, there is a constant need for high intensity lamps to light the plant. Also, the refinery production schedule varies widely during the four seasons of the year. Recently, the stores manager and the maintenance manager were discussing the service levels assigned to various spare parts. They were curious whether the 90-percent level assigned to the high-intensity sodium vapor lamps was adequate to prevent a stock out.

The Proposed Solution:

The storeroom manager decided to analyze the situation by applying the ROP formula (Equation 3-2, shown in Figure 3.15).

The Numbers:

Figure 3.16 shows the annual demand of 2,551 lamps broken down by months. Demand for the lamps varied from 4 per day in July to 10 per day in January, with average daily usage (Uavg) of 7 units. Because the lamps were not considered highly critical, a service level of 90 percent was considered acceptable. Routine lead time to replenish was 9 days. From the analysis the stocking levels were determined to be an ROP of 105 units and an EOQ of 159 units. The pattern for the maximum balance-on-hand and for the safety stock are shown in Figure 3.17. Notice that the time between reorders lengthens as the monthly demand decreases.

The Conclusion:

The analysis confirmed that the service level chosen was adequate to avoid stock outs if the safety stock was set at 42 units.

Plant Demand for High-Intensity Sodium Vapor Lamps

Demand during the year (Units per day)

Jan	Feb	Mar	Apr	May	June	July	Aug	Sept	Oct	Nov	Dec
10	9	8	7	6	5	4	5	6	7	8	9

Average usage/day (Uavg) = 7
Lead time (LT) = 9 days
Service level = 90%
Standard deviation of usage (S) = 11 units
Economic order quantity (EOQ) = 159
ROP = Expected usage during lead time + Safety stock

$$ROP = (Uavg)(LT) + (z)(\sqrt{LT})(s)$$

$$ROP = (7)(9) + (1.28)(\sqrt{9})(11)$$

$$ROP = 63 + 42 = 105$$

Where z is the factor from the normal distribution for a 90% service level

Figure 3.16 Ordering Pattern for Cyclic Use

The time between reorders lengthens as the monthly demand decreases. In this example, there are no stockouts at a 90% service level.

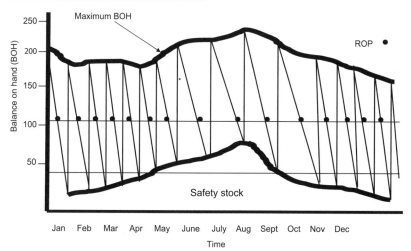

Figure 3.17 Ordering Pattern for Cyclic Use

3.5.2 Rarely-Used Items

In Chapter 1, we stated that the priority objective for rarely-used items was the "minimization of the total risk cost" (the sum of the costs of getting caught short plus the costs of getting caught long). Then in Chapter 2, we listed several different approaches for setting stock levels for slow-movers; two of these approaches were the square root of N and "just stock one of everything." For years it was common practice to use approaches like these, or simply to use the best guesses of maintenance, engineering, operating, and vendor personnel. Asking vendors how much to stock typically resulted in 20-40% overstocks. Asking maintenance resulted in 25-50% overstocks, with the normal cost and lead time biases thrown in. Asking materials management to guesstimate stocking levels frequently produced understocks.

Later, other statistical techniques such as "lumpy demand" (sometimes called "stuttering Poisson") were introduced to try to set stocking levels for rarely-used items. Because lumpy demand requires two or three demands in recent years, and because more than one-half of rarely-used key items (RUKIs) do not have that much usage, that technique was not very effective. Other techniques were suggested that used the number of months supply, or fraction of the prior years' demand. Still others ignored all the months of zero demand and based the stocking decision on the last actual month of usage. All of these techniques ignored at least several of the main inputs that determine proper stocking levels, such as criticality, number in service, and lead time. Even the advent of new computer hardware, online, real-time material systems, and complex mathematical algorithms have not done much to stop the buildup in inventory or the degradation in availability of spare parts, or both.

Then along came "risk-based assessment" as introduced by Inventory Solutions, Inc. (ISI) in 1984. ISI's stocking logic for rarely-used items was designed specifically for the risk-based assessment of that 85-90% of the operating and maintenance (O&M) spare parts for which other inventory management techniques have proven to be non-applicable, sub-optimal, useless, or counter-productive. Risk-based assessment is a different concept, principally because it "listens" to the reason for each rarely-used spare to be stocked, and then quantifies the probable on-hand inventory and probable backorder implications of alternative stock levels. Experience has also shown that non-key items do not warrant aggressive effort. They tend to be inexpensive and overstocked, comprising such a small portion of the inventory investment that the overstocking is of little concern to plant and finance management.

Conceptually, risk-based assessment calculations begin by calculating the probability that a specific part will require replacement on any given day. Some of the inputs for the calculation include historical usage and the implications of early failures after replenishment (see also Chapter 2). Multiplying that probability by the number of parts in service determines the overall likelihood that a request for an item from the storeroom will occur on any given day. Factoring this result by the lead time to replenish the spare part determines the probability of multiple demands occurring during the lead time to replenish. The final calculation in the process factors in the degree to which getting caught short is unacceptable. Fortunately, all this complex logic has been designed into an easy-to-use, risk-based decision support tool marketed by ISI as RUSL (Rarely Used Inventory Stocking Logic). The logic sequence is summarized below:

- Probability theory is used in the stocking model because the demand for the spare part can not be forecasted.
- Recent demand is used to help predict future failure rate.
- Recommended stocking level is limited to the period of the next lead time.
- Other modeling techniques are used to predict multiple random failures during the lead time.

Using the techniques of risk-based assessment first introduced in Chapter 2, the sensitivity of both the ROP and the average inventory to different levels of availability are shown in Figure 3.18. Notice the

Part cost = $1,000 Carrying cost = 20%   Cost of purchase order = $70.00

Availability	MIN/ROP	MAX	EOQ	Avg. ivny ($)
99.99	5	7	2	6,500
99.95	4	6	2	4,500
99.90	4	6	2	4,500
99.50	3	5	2	3,500
99.00	3	5	2	3,500
98.50	3	5	2	3,500
98.00	2	4	2	2,500
97.50	2	4	2	2,500
97.00	2	4	2	2,500
96.00	2	4	2	2,500
95.00	2	4	2	2,500
90.00	1	3	2	1,500
75.00	0	2	2	500

Availability combines first, the probability of having a failure of a part in service with second, having a replacement in the storeroom.

Figure 3.18 Stock Level Sensitivity to Availability

four-fold decrease in required average inventory as availability goes from 99.99 to 90.00. Figures 3.19 and 3.20 show the results plotted both against the minimum stocking level and the maximum stocking level.

Setting appropriate reorder points and reorder quantities is only part of the process of improving inventory balance. Once stocking parameters have been determined, older inappropriate values in the material management system must be purged and the new values inserted. When we update an incorrect stocking MIN and MAX, we are really setting a new inventory plan for the item. The process is simple in concept, but frequently hard to implement in practice, because plant management is often reluctant to make adjustments to long-standing stocking levels.

Setting the right stocking levels will not immediately lower inventory because the excess must either be used or be disposed of through

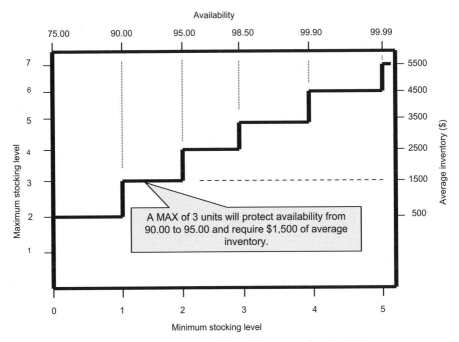

Figure 3.19 Stock Level Sensitivity to Availability

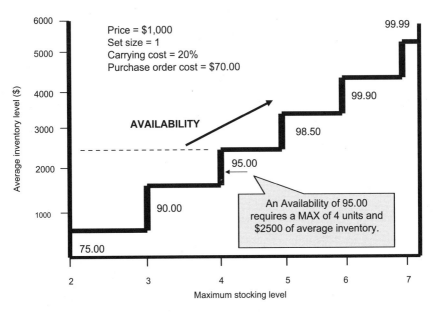

Figure 3.20 Stock Level Sensitivity to Availability

The table below shows how to fix your inventory plan by making adjustments to the reorder point and reorder quantity.

Stock code	Desired availability	Part cost ($)	Balance- on-hand (BOH)	August MIN	October MIN	August OQ	October OQ
05954	99.9	26.70	15	3	3	20	<u>24</u>
02224	97.0	5.42	634	200	6	500	35
16674	99.0	440.08	5	6	3	12	2
04453	97.0	553.39	2	12	1	61	1
09969	99.9	18.12	13	25	2	50	15
06763	99.9	554.17	17	6	2	12	1

Figure 3.21 Dynamic Inventory at Work

investment recovery (see Chapter 5). But having a plan for stocking each key spare part is essential if we are ultimately going to get a balanced inventory. There's an old saying: "If you don't know where your going, any road will get you there". The same is true if you don't have an inventory plan.

Figure 3.21 shows some typical adjustments to an inventory plan. Notice that, with only one exception (stock code 05954) the new MIN determined in October is lower than the August MIN. The same is true for the OQ (order quantity). Even though a rarely- used item is slow-moving, it does occasionally move. Over time, even a slight change in demand, price, or lead time can cause an adjustment to the MIN/MAX levels once in a while.

One of the greatest benefits of having a good decision support tool for setting stocking levels is the ability to perform various unique analyses. One analysis that comes to mind is how to convince maintenance managers when enough inventory is enough. Figure 3.22 shows a convincing way of doing that by calculating the incremental MAX for an item. In this example, to achieve a minimum availability of at least 99.91 (a high criticality item with 8-hours risk of running out), the

Price of part = $10,000   Needed availability = 99.91

MIN	MAX	EOQ	Availability	Avg. units	Average Invy ($)	Carry cost ($)	Incremental MAX ($)
3	4	1	99.65	3.01	30,059	6,012	6,666
4	5	1	99.94	4.00	40,020	8,004	34,348
5	6	1	99.99	5.00	50,014	10,003	199,872
6	7	1	99.99	6.00	60,013	12,003	Infinite

To calculate incremental MAX, divide the change in average inventory by the change in Availability for different stocking levels.

Figure 3.22 Incremental Maximum Can Tell You When Enough is Enough

MIN must be set at 4 units, resulting in an availability of 99.94. Stocking the MIN at 3 units only achieves an availability of 99.65, so we have increased the average inventory from $30,059 to $40,020 to get a 0.29 improvement in availability (99.65 to 99.94). Dividing the incremental average inventory increase ($9,961) by the incremental availability improvement (0.29), we get an incremental MAX of $34,348 to get a 1% improvement in availability.

If we set the MIN at 5 units, the incremental MAX becomes $199,880 to get the 1% improvement in availability. If we go up to a MIN of 6, we get no improvement in availability, and the incremental MAX becomes infinitely large. (Remember high school algebra where dividing a number by zero equals infinity!) Even the most die-hard maintenance managers will usually agree that when they spend a lot of money to get no benefit, "enough is enough."

Pareto's 80/20 Rule has been shown to apply to inventory as well as many other areas. Figure 3.23 shows Pareto's Rule at work for five spares from a recent plant analysis. Current and recommended stock levels are shown for each item and the dollar impact of changing the current MIN/EOQ to the recommended MIN/EOQ is shown in the last column. Even though the five items represent only 2% of the SKUs in the analysis, they make up 21% of the dollar impact when the changes are made. This analysis is typical for most rarely-used items: 10-20%

By applying Pareto's Rule, we see that a small percentage of slow-moving inventory can account for a large percentage of savings. These five items are only 2% of the items, but make up 21% of the impact dollars.

Part No.	Lead time (Weeks)	Current stock level			Recommended level			Impact ($)	
		MIN	MAX	EOQ	MIN	MAX	EOQ	Per item	Cumulative
070011	9	3	28	25	2	3	1	16,962	16,962
027002	9	12	42	30	3	9	3 S	5,422	22,384
033417	6	6	18	12	1	2	1	4,189	26,573
076655	8	5	7	2	3	4	1	3,303	29,876
044545	7	2	4	2	1	2	1	2,168	32,044

S = Issued in set of 3

Figure 3.23 Pareto's Rule at Work

A 1957 nomograph determines the reorder point for items as a function of the *average number of demands during the lead time*. A comparison with a modern algorithm is shown below.

Results based On 3 years of usage with 3 demands

Average lead time demand	Nomograph reorder point	Modern algorithm reorder point
0.25	2	3
0.50	3	4
1.00	5	5
2.00	7	8

The modern algorithm is more conservative in most cases

Figure 3.24 Some of the "Old Ways" are Still Good

of the items account for 75-80% of the impact. Notice also that, for part 027002, the item was determined to be a set of 3. Therefore, the EOQ and the MAX are a multiple of 3 units.

Much progress has occurred over the last fifty years to improve the management of production plant inventory. Faster computers, better

What if your supplier can not ship your full replenishment? The shape of the replenishment pattern depends on the demand for the item

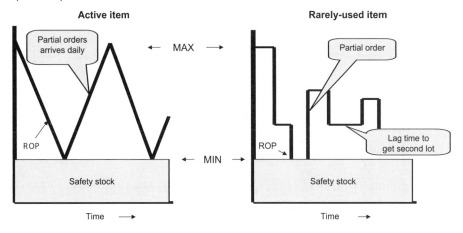

Figure 3.25 Replenishment May Not Arrive in One Batch

data bases, and more sophisticated software have all helped. But sometimes we forget that some of the "old stuff" developed years ago was pretty good for doing the stocking job intended. A case in point is a 1957 *National Institute of Management* nomograph about determining the reorder point for items as a function of the average number of demands during the lead time. Figure 3.24 shows some comparative results over a range of different demands during the lead time. Notice that, in all cases, the difference in order point recommended by the nomograph and the modern algorithm is one unit or less of inventory, with the modern algorithm being slightly more conservative by recommending a higher ROP.

Two final points need to be made before concluding this chapter. First, when a purchase order is issued for a replenishment, the order may not be completely filled by the supplier, resulting in partial delivery (see Figure 3.25). Two cases are shown, one for an active item supplied by a distributor and the other for a slow-mover. In both cases, the replenishment arrives over a period of time rather than a single shipment. The time span to complete the full replenishment is likely to dependent on the distributor's ability to replenish his stock, which in turn depends on a manufacturer somewhere to produce the item.

Second, it is not necessary to stock everything, especially for rarely-used items. Three factors, taken together, usually determine whether or not it is necessary to stock an item:

- **The risk** you are willing to take against running out. Can you go without the spare part for a few days?
- **The lead time to replenish.** Can you get it in a week or less?
- **The average annual usage.** Do you need it once per year or less?

If all of the above criteria can be met, the odds are that the part does not need to be stock at all, but only replenished when there is a demand for the spare part at the storeroom counter.

Plug valve

CHAPTER 4

SETTING THE REORDER QUANTITY

4.1 WHAT THE READER WILL LEARN FROM THIS CHAPTER:

- The key factors affecting the reorder quantity (ROQ)
- The standard economic order quantity (EOQ) formula
- How EOQ is calculated with price discounts
- Several concepts for setting replenishment quantity
- Unique solutions to everyday problems

4.2 FACTORS INFLUENCING THE EOQ

In Chapter 3 we learned about setting the reorder point. In this chapter we will discuss the companion issue of how to set the reorder quantity – the amount that should be purchased when the decision has been made to buy.

Figure 4.1a shows the standard textbook equation for calculating the reorder quantity. Four major factors influence the calculation: 1) the annual **demand** for the item, 2) the **cost to issue a purchase order,** 3) the **carrying cost factor**, and 4) the **price of the item**. Of the four, only demand is a common factor in setting both the reorder point and the reorder quantity. Because the terms within the square root are a quotient, any combination of values that causes the numerator to increase, or the denominator to decrease, will cause the EOQ to increase.

Sample calculations for both rarely-used and active items are shown in Figures 4.1b and 4.1c, respectively. Our experience in setting stocking levels for tens of millions of rarely-used spare parts indicates that the formula-derived EOQ will be one unit about 80 percent of the

$$EOQ = \sqrt{\frac{2RS}{KC}} \qquad (Eq\ 4\text{-}1)$$

Annual demand → 2 R S ← Cost of issuing a purchase order

Carrying cost → K C ← Item price

a

Annual demand (R) = 1
Cost to issue P.O. (S) = $50
Carry cost (K) = 20 %
Item cost (C) = $500

$$EOQ = \sqrt{\frac{2 * 1 * \$50}{0.20 * \$500}} = 1$$

b

R = 3,650
S = $50
K = 20%
C = $10

$$EOQ = \sqrt{\frac{2 * 3{,}650 * \$50}{0.20 * \$10}} = 427$$

c

Figure 4.1 Calculating EOQ for Active Stock

time because of the low annual demand for most items. Therefore, if you have to guess what the EOQ would be for a rarely-used item, you would be correct 80 percent of the time by choosing one unit. In the following subsections, we will discuss each of the four factors in more detail.

4.2.1 Annual Demand

The annual demand for an item is usually obtained by looking at the usage history of the item captured in the plant material management system. In most cases, the value shown is the total demand, or issues from the storeroom, without consideration of whether the need was to replace a failed part, for a scheduled overhaul, or for some other reason. When setting stocking levels for safety stock, which is what we are mainly concerned about for rarely-used items, we prefer to only consider that part of total demand that relates to part failure. However, most plant material systems do not distinguish between failure demand and other demand. As a result, we typically have to include the total demand when applying the EOQ formula, even though it tends to inflate the reorder quantity. It is suggested that plant record systems be modified to sort issues into different categories such as: 1) failures, 2) overhauls, 3) capital projects, and 4) other uses. After tracking these categories for two or three years, the accuracy of both the reorder point and reorder quantity calculations will improve.

One final point regarding the annual demand component in the EOQ formula. Because it is common to have many years in which a rarely-used spare is not used at all, entering a zero in the EOQ equation for **R** doesn't work. The numerator of the equation would become zero, and the square root of zero is zero. Therefore, to apply the equation at all, the value of **R** must be positive. Entering a value of 0.1 for the annual demand is usually effective in calculating a valid EOQ. Decision support software for rarely-used spares tends to inflate actual usage (even zero usage) in order to be conservative; therefore, adjusted demand for slow-moving items is always positive.

4.2.2. Cost of Issuing a Purchase Order

Determining the cost of processing a purchase order can be tricky. Should the calculation include both blanket orders and stand-alone purchases? Should it factor-in obsolescence? What about expediting costs? All of these considerations will affect the final value used for S in the EOQ equation.

To arrive at a value for **S**, an analysis of the last year or two of purchasing history is necessary. Most companies tend to include in that analysis the following cost components: 1) purchasing department overhead for managers, buyers, and clerical personnel, 2) expediting, 3) receipt and inspection, 4) accounts payable, 5) freight and shipping, 6) computer expense, and 7) obsolescence (see Figure 4.2).

Cost of ordering inventory:

Management, clerical, and computer costs to create
  purchase order
- Receipt and inspection costs
- Expediting costs
- Cost to process accounts payable
- Quantity discounts, freight consolidation breaks

Cost of carrying inventory:

- Property taxes and insurance
- Depreciation and obsolescence
- Warehouse overhead
- Interest on money tied up in storeroom stocks
- Storeroom labor costs for managing inventory

Figure 4.2 Factors Influencing the EOQ

Once the annual cost for these components are determined, the sum is divided by the number of purchase orders processed during the year to arrive at a dollar value per average purchase order. In most cases, blanket orders are not included in the computation because they use only a small part of the above cost components.

Values for **S** can vary widely depending on the industry. For most industries, a range from $50 to $100 is normal, but values of $200 to $300 are common for nuclear power plants where the higher costs of safety certification and inspection come into play. Frequently, rather than go through the effort of calculating the value for their plant, purchasing managers just check around their industry and see what others are using as a cost of processing orders. Unfortunately, many plants don't even bother to do that; instead, they just don't use the EOQ formula at all.

4.2.3 Carrying Cost Factor

Determining the value to use for the carrying cost factor **K** in the EOQ formula can be subjective. Usually the value used is the one determined by the chief financial officer (CFO) or the comptroller, and is used company-wide. A number of factors may or may not be used in determining the factor including: 1) local taxes on inventory, 2) warehouse and inventory insurance, 3) storeroom depreciation, 4) obsolescence of inventory, 5) warehouse salaries and overhead, and 6) interest costs on the money tied up in storeroom stocks (see Figure 4.2).

Again, values range widely depending on company and accounting policy. Many companies use only the cost of borrowing, which can vary from 5% to 15%; others use the all-in cost, which can easily range from 20% to 30%. The cost of money tied up in inventory should certainly be included because those dollars are borrowed, either from the bank or the stockholders.

4.2.4 The Item Price

You would think that the price of the item you are buying should be the least controversial of all the formula factors. In most cases, it is. Yet other factors must still be considered when selecting the value for **C** in the EOQ equation: 1) whether to use the current vendor price or our average unit price in the material system, 2) whether to include or exclude any discounts, and 3) whether to include or exclude freight costs and sales taxes. These decisions are usually made by the financial people; it probably doesn't matter too much what they include or exclude as long as the decision is consistent from year to year.

4.2.5 How the Input Parameters Influence the EOQ

Figure 4.3 shows some calculated values for EOQ as demand, purchase order cost, carrying cost, and part price vary. For a low-demand item, the EOQ ranges from 1 to 8 units as the price varies 10-fold and the cost to process a purchase order varies 3-fold. For a high-demand item, EOQs range from 10 to 45 units. Many inventory management systems include an automatic calculation of EOQ as part of the ordering process. If your system doesn't, it's not a difficult job to have a stand-alone calculator programmed into your system. Let's look at several case studies regarding EOQ.

Demand (R)	Purchase order cost (S)	Carrying cost (K)	Part price (C)	EOQ (Rounded up)
1	$ 50.00	0.25	$ 500.00	1
1	150.00	0.25	500.00	2
1	50.00	0.25	50.00	3
1	150.00	0.10	50.00	8
50	50.00	0.10	500.00	10
50	150.00	0.10	500.00	18
50	150.00	0.25	50.00	35
50	50.00	0.10	50.00	45

Figure 4.3 How Parameters Affect EOQ

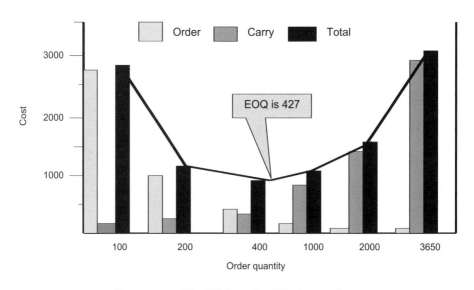

Figure 4.4 The EOQ is the Minimum Cost

CASE STUDY 4-1 AJAX HYDRAULIC

The Situation:

AJAX Hydraulic is a small eastern U.S. company manufacturing hydraulic actuators and other remote-driven components for the chemical and petrochemical industries. Most of the parts for the components are purchased from numerous vendors and are then hand-assembled into the final product. The company is extremely cost and quality conscious; it enjoys a good reputation in the industries it serves. Minimizing cost is a high priority.

According to the plant purchasing manager, "Because we buy so much of what goes into our product, we keep a close eye on every dollar we spend. I probably spend over one-half of my time negotiating better prices and terms from our suppliers. I'm a strong advocate of using the EOQ formula to determine the right quantity to buy each time. We have a sharp analyst in our department who evaluates each major purchase to determine what the right buy amount is. Let me show you an analysis she did yesterday."

The Proposed Solution:

The analyst used the standard EOQ formula to calculate the optimum purchase amount for the next quarter's production requirements for hasp pins used on their Model 97G remote actuator. In addition to running the optimum EOQ number, she also plotted the EOQ over a range of order sizes, as shown in Figure 4.4.

The Numbers:

Production schedules for the next quarter called for 3,650 pins. The value for S was determined by a previous analysis to be $50/order, and the carrying cost factor was set at 20 percent by the plant comptroller. Once the calculated EOQ was determined to be 427 units, a purchasing schedule was set up on that basis. Figure 4.4 also shows the impact on EOQ of both the order cost and the carrying cost. The order cost goes down as the order quantity increases, whereas carrying cost increases.

The Conclusion:

Based on the negotiated price of $10.00 for the pins, a purchasing schedule was set up to buy the pins in lots of 427.

CASE STUDY 4-2: AJAX DISTRIBUTORS

The Situation:

AJAX Distributors is a small midwest distributor of electrical parts and components to the general industry. It distributes a variety of parts ranging from fractional horsepower motors to a variety of switching gear. One component it sells regularly is a heavy-duty industrial fuse costing about $50 from the manufacturer. The purchasing manager was an expert on freight costs and typically negotiated freight discount prices regularly.

He put his thoughts this way, "I have a 'show me' mentality when it comes to using some of those high-powered formulas for calculating how much I should buy. I like to calculate the buy-amount the hard way occasionally, just to prove to myself that the standard EOQ formula is okay to use routinely. I just finished a comparison for our next fuse order."

The Proposed Solution:

The purchasing manager gathered the necessary inputs to determine the buy amount, as shown in Fig.4.5. He excluded the freight cost from the administrative cost because he wanted to break freight cost out separately. Order quantities from 50 to 1000 units were evaluated.

The Numbers:

To buy the total annual demand of 1,000 units in one lot resulted in a total annual cost of $5,650 ($50 for administrative cost, $100 for freight, $500 for the carrying cost on the safety stock amount, and $5,000 for the carrying cost on the average number of cyclic units). Spreading the order over 20 lots of 50 each brought the annual total cost down to $1,950. The optimum for the six scenarios studied turned out to be 10 lots of 100 units each. The purchasing manager also determined that if he added $20 to the cost of processing a purchase order (to factor in the freight cost per order), the standard EOQ formula generated an EOQ of 118 units.

The Conclusion:

The purchasing manager decided to rely on the standard EOQ formula to calculate the order lot size rather than spend the extra time and effort to do the calculation the long way.

Inputs:

Demand = 1,000 units/yr
Weight/unit = 5 lbs
Admin cost = $50/order
(Excludes freight)

Carrying cost = 20%
MIN = 50 units (Safety stock)

Unit cost = $50.00
Freight rates: 2 cent/lb > 4,000 lb
3 cent/Lb >1,000 < 4,000 lb
4 cent/lb < 1,000 lb

Order quantity	Orders per year	Admin cost ($)	Freight cost ($)	Carrying cost ($) [Safety stock]	Carrying cost ($) [Cycle stock]	Total cost per year ($)
1000	1	50	100	500	5,000	5,650
500	2	100	150	500	2,500	3,250
250	4	200	150	500	1,250	2,100
200	5	250	150	500	1,000	1,900
100	10	500	200	500	500	1,700
50	20	1,000	200	500	250	1,950

☐ Optimum order quantity   Note: Adding $20/order to admin cost = 118 units by EOQ formula

Figure 4.5 Calculating EOQ the Hard Way

CASE STUDY 4-3: AJAX ENTERPRISES

The Situation:

AJAX Enterprises is a $290-million-per-year supplier of parts and components to the aerospace, automobile, and electrical equipment industries. Like most of its competitors, it both makes and buys a substantial amount of the parts and pieces that go into its components. The plant manager is very cost conscious and constantly rides herd on the production and purchasing managers to do everything they can to keep costs down. He was especially disturbed recently when he was sent an approval request to purchase a large lot of parts for one of their units. "I had a gut feel that the size of the order quantity was way too high, and I told the purchasing manager to rework the numbers, and apply the EOQ training from a seminar held earlier this year."

The Proposed Solution:

The purchasing manager was asked to do an analysis that would show the consequences of not ordering the EOQ amount.

The Numbers:

Figure 4.6 shows the results of the computation. The black bars show the percentage by which the Order Cost is over the EOQ Order Cost when the lot size was varied from the EOQ lot size of 427 Units. The gray bars show

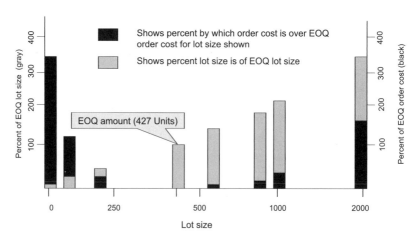

Figure 4.6 The Cost of NOT ORDERING the EOQ Can be Substantial

what percent each chosen lot size was compared to the EOQ lot size.

The Conclusion:

The plant manager made a poster of the chart and attached it to the wall in each buyer's office. His parting words were, "buy the EOQ amount, unless you can convince me otherwise!"

 ## CASE STUDY 4-4 AJAX METALS

The Situation:

AJAX Metals is a southeastern plant engaged in the machining of heavy steel castings into final shapes for sale to the utility and construction industries. A major purchase for the company is abrasive grinding wheels and cutting tools. In reviewing some purchase orders for new spares for some recently-installed equipment, the operations manager noticed a major inconsistency in the EOQ suggested by two different decision support codes the company had acquired. "I don't get it. We have these two codes that are suppose to recommend the MIN/MAX for the new spares, but the numbers differ widely. Which set of numbers should we use to buy?"

The Proposed Solution:

The supply chain manager looked at the MIN/MAX values from the two codes (labeled Code A and Code B rather than by the trade name so as not to bias the conclusions when she presented the results to the operations manager). She first compared the recommended MINs for the sixteen items and then compared the recommended MAXs.

The Numbers:

Here's what she found regarding the MINs:
- *Code A recommended a MIN higher than Code B seven times*
- *Code A and Code B recommended the same MIN four times*
- *Code A recommended a MIN lower than Code B five times*

She concluded that the comparison of the recommended MINs by the two codes was a wash!

When the supply-chain manager looked at the MAX values recommended by the two codes, she found big differences. Code A used the standard formula to calculate the EOQ, then added that value to the MIN to generate the MAX. Because five of the sixteen items had costs of less than $10, Code A calculated a fairly large EOQ for these items. Code B did not appear to use the EOQ formula at all. In fact, four of the recommendations using Code B had an EOQ of zero (the MIN was the same as the MAX). The supply chain manager tried one more thing. She took the EOQ amount calculated by Code A and added the value to the Code B MIN to calculate an adjusted Code B MAX.

The Conclusion:

Comparing the Code A order quantity to the adjusted Code B order quantity, the total buy using Code B was $4,109 more than Code A. The supply-chain manager recommended that the operations manager use the Code A numbers for the purchase, and Code A for future new spares purchases.

Tubing

$$EOQ = \sqrt{\frac{2\,R\,S}{K\,C} \times \frac{P}{(P\text{-}D)}} \qquad \text{(Eq 4-2)}$$

The following parameters also apply to Figure 4.8:

P	= Production rate	=	75 Units per day
R,D	= Normal demand rate	=	25 Units per day
KC	= Carrying cost	=	$ 0.03 per unit
S	= Order cost	=	$75.00

Resulting in an economic run size = 433 units

Figure 4.7 The EOQ Formula Can Accommodate the Production Schedule

4.3 THE PRODUCTION SCHEDULE AND THE EOQ

Many times when a plant expects to order a large quantity of an item over an extended period, it is advisable to coordinate the delivery schedule with the part manufacturer. This coordination can be accomplished by adjusting the standard EOQ formula to include the expected production rate of the supplier. Equation 4-2 (see Figure 4.7) shows how this adjustment is made. In this example, the EOQ using the standard formula would be 354 units without adjusting for the production rate. When the production rate is factored, the EOQ increases to 433. The following case study shows how the customer and the supplier worked together to handle the order.

CASE STUDY 4-5 AJAX MANUFACTURING

The Situation:

AJAX Manufacturing and one of its long-standing customers were discussing a recent purchase order. The customer had recently installed a new production line and would need periodic shipments of one of AJAX's components. Because the customer expected to run the new line in a batch mode, it wanted the component shipments to arrive over a three-week period for each production campaign.

The Proposed Solution:

AJAX's sales manager worked out a production/ shipping schedule where components would be manufactured and shipped simultaneously.

The Numbers:

Maximum daily production of the component at the AJAX factory was 75 units per day. Because some of the equipment used to manufacture the component was also used for other products, a two-day set-up period (Set Up Lead Time, or SLT) was required. It was agreed that AJAX would manufacture the component at 75 units per day and ship at a rate of 25 units per day. Subsequent runs would follow the same production schedule unless modifications were suggested by the customer. AJAX was able to include in its price the carrying cost associated with the buildup of finished goods inventory, as shown in Figure 4.8.

The Conclusion:

The customer's purchasing manager stated, "We like working with AJAX. They always are willing to accommodate our needs, and I especially like it that they don't charge us extra for their set-up time."

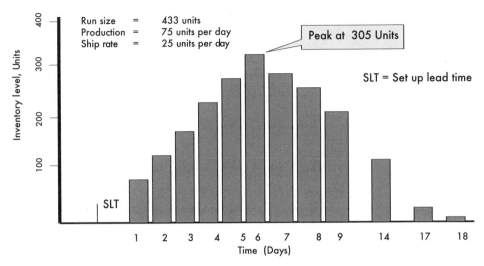

Figure 4.8 Accommodating the Economic Run Size

4.4   SETTING THE EOQ WITH PRICE DISCOUNTS

It is fairly common to negotiate price discounts as the size of an order increases. Discounts can be an incentive for both the buyer and the supplier. Buyers get a better price if they buy more whereas suppliers get a larger firm order that frequently allows for more efficient and cost-effective production. Because of the carrying cost of larger orders over time, there is usually a limit on the size of an order that is economical to the buyer. To determine the optimum order size, it is necessary to consider three factors: 1) the cost of placing orders, **S** in the EOQ formula, 2) the carrying cost factor, **K**, and 3) the product cost, C. The following case study for AJAX Electronics shows how discounts can be handled.

CASE STUDY 4-6 AJAX ELECTRONIC

The Situation:

AJAX Electronics is a medium-size manufacturer of control instrumentation for the chemical, power, and refinery industries. Its only factory is in Indiana from which it ships finished components both domestically and to the international market. The purchasing manager had recently negotiated some discount prices from a supplier for a 20-amp fuse used in one of their instruments and now needed to determine what total order quantity to place. "The discounts we received were better than what we got last year. The lowest price was $8.50 per unit, but we would need to buy at least 2,500 units to get that price (See Figure 4.9).

The Proposed Solution:

The purchasing manager, who needed to consider the trade-off between buying several smaller lots or one large lot, decided to develop a spreadsheet of the costs of buying lots of 400, 1000, 2000, and 3650 units (one years supply).

The Numbers:

Figure 4.10 shows the results of the analysis. Order costs went up almost 10-fold as the order size went from 3,650 to 400 units. Similarly, the carrying cost increased 8-fold as the order size increased from 400 to 3,650 units. However, the increase in carrying cost was more than offset by the lower unit price to place an order for a full year's production.

The Conclusion:

The purchasing manager placed an order for 3,650 units after negotiating a further 50-cent discount on the next order of similar size.

Order quantity (OQ)	Price per item
0 < OQ < 500	$ 10.00
500 < OQ < 1500	$ 9.50
1500 < OQ < 2500	$ 9.00
2500 < OQ < 4000	$ 8.50

Figure 4.9 Setting the EOQ with Price Discounts

Planned order quantity and price

Cost item	400 $10.00	1000 $9.50	2000 $9.00	3650 $8.50
Order cost ($)	456	183	91	50
Carrying cost ($)	400	900	1,700	3,102
Product cost ($)	36,500	34,675	32,850	31,025
Total cost ($)	37,356	35,758	34,641	34,177

Annual demand equal 3,650 Units   Order cost = $50   Carrying cost factor =0.2

Figure 4.10 Optimum EOQ with Price Discounts

4.4   DIFFERENT CONCEPTS FOR MANAGING REPLENISHMENTS

Although ordering replenishments using the EOQ formula is pretty much the standard approach, other concepts can be used to determine when and how much to buy. This section will discuss several of these accepted practices. Material in this section is based upon *Supply Chain Strategy* by Edward Frazelle (with permission).

4.4.1 The Traditional EOQ Concept

Figure 4.11 shows a graph of inventory level versus time when a replenishment order equal to the EOQ is placed when the reorder point in the material system is reached (when the balance-on-hand hits the ROP). Depending on the demand for the item, the balance-on-hand in the storeroom can dip into safety stock before the replenishment order is received. Under this concept, the order quantity is a constant amount from order to order unless price or other factors change the calculated EOQ amount. The advantages of this concept are: 1) it is the most commonly used, 2) it is easy to understand, and 3) it uses a fixed-order quantity.

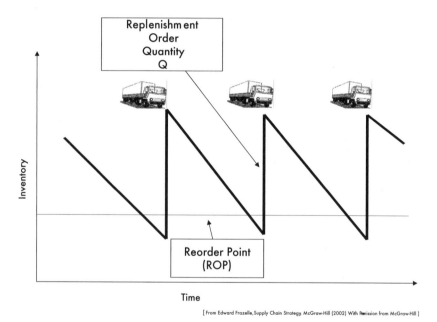

[From Edward Frazelle, Supply Chain Strategy, McGraw-Hill (2002) With Permission from McGraw-Hill]

Figure 4.11 The Traditional Concept for Setting Replenishments

4.4.2 Ordering a Variable Amount Instead of the EOQ

A second concept for replenishment, in which a variable amount is ordered when the reorder point is reached, is shown in Figure 4.12. Under this concept, it is necessary to estimate the expected demand

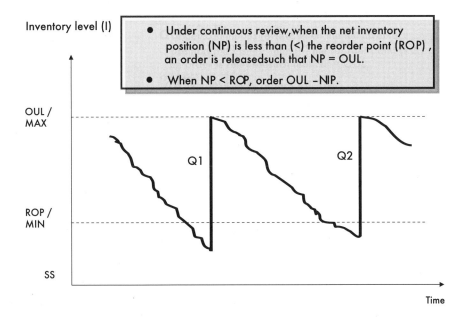

Figure 4.12 Ordering at Preset Times to Set Replenishments

for the items during the replenishment lead time, then determine the order size each time to bring the balance up to the "order-up-to-level" (OUL) or MAX. In the example shown, order size Q2 is smaller than Q1. The principal advantage of this concept is that the MAX will not be exceeded.

4.4.3   Ordering A Variable Amount At Preset Times

A third concept for replenishment is to use preset review times to order a variable amount that restore balances up to the MAX. The principal advantages of this approach are: 1) there is less chance of stock outs, and 2) there is an opportunity to adjust the reorder quantity more often. Offsetting disadvantages include the need for more frequent reviews and the potential for higher carry cost. In the example shown in Figure 4.13, each replenishment order is placed midway during a two lead-time period. Notice the large differences in the size of the replacement order between the fourth time period (t) and the sixth time period.

4.4.4 Ordering A Variable Amount Depending on the Balance

The final concept for setting reorder size is perhaps the most complicated and difficult to understand (see Figure 4.14). Here, the approach is to place the replenishment order *only if* the balance-on-hand is at or below the reorder point during the review period. The size of the variable replenishment quantity is set to equal the OUL/MAX minus the expected balance-on-hand (I) when the order is placed. Usage during the period of replenishment tends to keep the inventory level from reaching the OUL/MAX. The frequency of ordering also can vary widely. The main advantage of this approach is it can result in fewer orders than the other concepts.

Figure 4.15 summarizes the advantages and disadvantages of the four concepts.

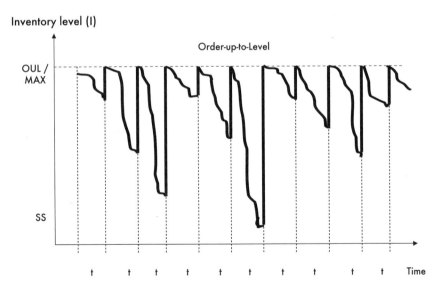

Figure 4.13 Ordering at Preset Times to Set Replenishments

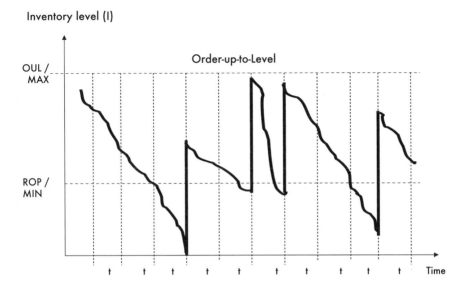

Figure 4.14 Set Replenishments Depending on the Balance-on-hand

Control Policy	Advantages	Disadvantages	Notes
ROP/EOQ	Simple system, fixed order quantity.	Can't cope with big swings.	EOQ computation may be unreliable
ROP/OUL	Best ROP/OUL = Best ROP/EOQ.	Heavy computational effort not justified except for At items.	Most popular policy but with arbitrary parameters.
RTP/OUL	Coordination of replenishment of related items. Regular opportunity to adjust the OUL.	Carrying costs are higher than in continuous review systems.	Good when demand pattern is changing regularly with time.
RTP/ROP	Overall less cost control policy.	High computational Effort. Difficult to understand.	Not justified for B or C items.

Figure 4.15 The Different Replenishment Concepts

The concept of ordering a amount to replenish to the MAX during preset review times:

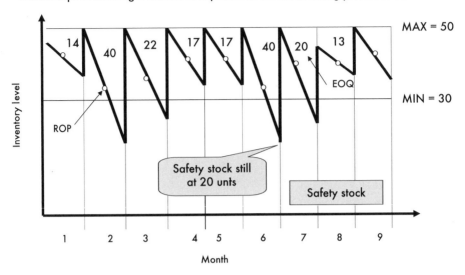

Figure 4.16 AJAX Lubricants Ordering Pattern

Spare Parts Rack

CASE STUDY 4-7 AJAX LUBICANTS

The Situation:
AJAX Lubricants provides a variety of lubricating oils, greases, and pastes used by special machining shops around the country. The base for their lubricants is a highly-refined mineral oil to which they add a mix of additives and grinding power depending on the specific grinding application intended. Because their orders are cyclic, depending on the work load of their customers, they prefer to purchase additives only as needed to fill orders. Some additives have limited shelf-life after mixing with the base; prolonged storage can impact on the quality of the final product. The lead buyer for the company put it this way, "We know it's a bit more work to review our additive ordering requirements monthly, but we think it keeps our stocks in line better with our shipping requirements to our customers."

The Proposed Solution:
AJAX elected to base their replenishment orders on a monthly review conducted the 15th of each month, when they estimated usage during the replenishment lead time and bought only the number of drums of additive necessary to bring inventory up to the MAX of 50 drums.

The Numbers:
Figure 4.16 shows the ordering pattern for AJAX over the last 9 months. Order size varied over the period from a low of 13 drums to a high of 40. At no time did the safety stock, which is always rotated on a first-in, first-out basis, ever drop below 20 drums. Because of the shelf-life issue, the company felt the more frequent ordering pattern was justified. Only in month 7 did the forecast of demand during the lead time result in not bringing stocks up to the MAX after replenishment.

The Conclusion:
AJAX continues to use this approach for reordering stock.

4.4.5 Some More Concepts for Reviewing Replenishments

A number of additional concepts for deciding when and how much to order are covered in this section. These options are from concepts used by or tried by clients in the past. Sample data showing how each concept works are documented in the following case study.

Although it is best to use a decision support tool to manage stocking decisions, those who prefer self-management may want to try these approaches:

Decision driver:	When to replenish:	Amount to replenish:
Balance-on-hand	Order when balance-on-hand equals the MIN (traditional approach)	Order EOQ amount or amount to get to MAX
Balance-on-hand	Order when balance equals 2X safety stock	Amount to get to MAX
Timing	Order first day each month	Amount to get to MAX
Timing	Order every lead time period	Amount to get to MAX
Usage	Order when 3-month moving average of usage equals +/- 1.2 times previous period usage	Amount to get to MAX

Figure 4.17 Options for Ordering ReplenishmentsFigure

Conditions: Review if Balance-on-Hand < 2x MIN
Active item
Order an amount to replenish to MAX

Stock No	BOH	MIN	MAX	Review (Y/N)	Decision
49874	798	400	900	Yes	Order 112
63382	405	500	1000	Yes	Order 595
11353	725	1000	2000	Yes	Order 1,275
43800	2	2	6	Yes	Order 4
08998	51	25	100	No	No order
96866	12	10	12	Yes	No order, Balance = MAX
74744	61	110	210	Yes	Order 147

Figure 4.18 Reordering Based on Balance-on-hand

CASE STUDY 4-8: AJAX TECHNOLOGIES

The Situation:

AJAX Technologies is a fast growing manufacturer of innovative components to the aircraft, automobile, and trucking industries. The company was founded in 1997 and has been growing rapidly, having added a major expansion to the factory in 2002. At a recent weekly staff meeting, the manufacturing manager was lamenting a recent conversation with the Vice President, "The Chief is getting concerned about our lack of a decision support tool to help us manage our spares replenishments. I said we've been doing okay using our 'best judgment,' but maybe we better see what's out there." The supply-chain manager added, "I don't think we need some fancy tool yet. Let me have Lee contact some other people and see if they are using anything better than we are."

The Proposed Solution:

Lee in materials management was assigned to contact several of their suppliers and see what they were using, in order to determine when and how much to replenish.

The Numbers:

Figure 4.17 summarizes five concepts for deciding when and how much to order. All were considered appropriate only for active items. The traditional concept of ordering the EOQ when the MIN was reached was used by many companies, although several ordered an amount to bring the balance back to the MAX instead of the EOQ. Other concepts used by several suppliers were based on ordering at a prescribed time (first of each month), or tracking usage fluctuations to decide when and how much to buy. Figures 4.18 through 4.21 show sample data for how each concept would work in practice.

At the next staff meeting, the supply-chain manager commented on the four examples as follows:

• *The concept of initiating an order when the BOH falls below two times the MIN has promise. About one-third of the time there is no need to order.*
• *The concept of ordering an amount each month equal to what was used in the prior month also has promise, especially if you order back up to the MIN/ MAX mid-point.*
• *The third concept of reviewing every lead time may have problems. In this example, it led to a stock out. That could be okay, however, if the part is not critical.*
• *The last concept of adjusting the MAX each month could have merit also, but it looks like it would be more difficult to manage the numbers.*

The Conclusion:

Ten spares having a price between $300 and $500 were selected. It was decided to review them using each concept for three months, then report back with the findings at the first staff meeting next quarter.

Conditions: Lead time to replenish = 4weeks   Current MIN = 50
Review monthly   Current MAX = 100
Active item
Order quantity to replenish amount used in prior month

	J	F	M	A	M	J	J	A	S	O	N	D
Balance first day of month	75	67	65	75	64	57	75	57	75	9	75	63
Amount ordered (Units)	0	8	10	0	11	18	0	18	0	66	0	12
Amount used each month	8	10	0	11	18	0	18	0	66	0	12	9

Figure 4.19 Reordering Based on Timing (Monthly)

Conditions: Lead time to replenish = 8 weeks   Current MIN = 50
Review every 2 months   Current MAX = 100
Active item
Order amount to bring balance up to MIN/MAX midpoint

	J	F	M	A	M	J	J	A	S	O	N	D
Balance first Day of month	42	34	57	57	64	46	57	57	75	9	9	-3
Amount ordered (Units)	33	-	18	-	11	-	18	-	0	-	66	-
Amount used each month	8	10	0	11	18	0	18	0	66	0	12	9
Balance when order received	-	24	-	46	-	46	-	57	-	0	-	-12

Figure 4.20 Reordering Based on Timing (Leadtime)

Conditions: Lead time to replenish = 4 weeks Current MIN = 40
Review monthly   Current MAX = 200
Active item
Adjust MAX by 1/2 difference from prior period
Order amount to bring balance up to MAX

	J	F	M	A	M	J	J	A	S	O	N	D
3-Month moving average Avgbalance	106	244	201	106	16	8	9	10	95	177	236	200
Review if: Avg < MIN	No	No	No	No	Yes	Yes	Yes	Yes	No	No	No	No
Avg > MAX	No	Yes	Yes	No	No	No	No	No	No	No	Yes	No
Adjustment to MAX	-	+69	-21	-48	-45	-4	+1	+1	+43	+41	+30	-18
New MAX	-	269	248	200	165	161	162	163	206	247	277	259
Quantity ordered (Units)	-	25	47	94	149	153	153	153	111	70	41	59

Figure 4.21 Reordering Based on Usage

CHAPTER 5

DETERMINING WHAT'S EXCESS

5.1 WHAT THE READER SHOULD LEARN FROM THIS CHAPTER

- How to determine gross overstocks
- Typical overstocks for various industries
- How fast overstocks should work off through attrition
- A balanced inventory plan
- Four ways to help optimize your inventory investment

5.2 DETERMINING GROSS OVERSTOCKS BY THE EXTREME TEST

Usually before a company is willing to invest its financial and human resources to initiate an inventory optimization program, it must be convinced it has an inventory imbalance problem. The problem is that most companies don't want to spend any resource to be convinced. Here's where the Extreme Test method developed by Inventory Solutions, Inc., comes into play.

In Chapter 3 we talked about the key inputs needed to set stocking levels for slow-moving spares. We showed that the criticality and lead time to replenish are two key input parameters, with demand and number of parts in service also important. One way to arrive at a reliable estimate of overstocks is to assume conservative values for criticality, lead time, and number in service, and then use the available demand history from the material management system. Figure 5.1 shows an overview of the method.

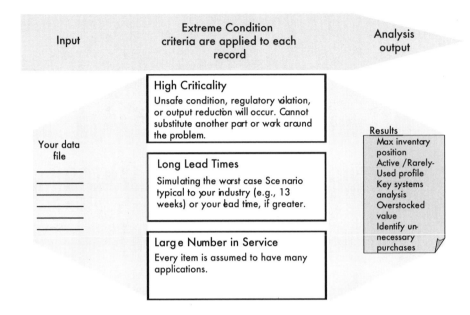

Figure 5.1 Determining Gross Overstocks

Typical extreme test assumptions used are: 1) all parts are highly critical and require a minimum of 99.9 percent availability (8-hours risk of not having the part during the year), 2) a long lead time (typically 39 weeks for the power industry and either 13 or 26 weeks for other industries), and 3) a large number-in-service for each item (usually 15). Output from the extreme test consists of: 1) an estimate of the maximum stock position needed to support the 99.9 percent availability, 2) an estimate of the amount of overstocks, and 3) a projection of unnecessary purchases. When using the extreme test method, we are intentionally setting a very high required level of inventory for each item using the conservative parameters. If the actual stock level for the item is greater than the calculated level, the odds are about 95-plus percent the item is overstocked.

Figure 5.2 shows a graphical representation of the extreme test. Graph 1 shows the current inventory value for an item as the shaded area impacted by criticality, lead time, and number in service. Graph

Gross overstocks are determined using extreme condtions. The value of overstocked items can significantly increase when applying actual criticalities and lead times.

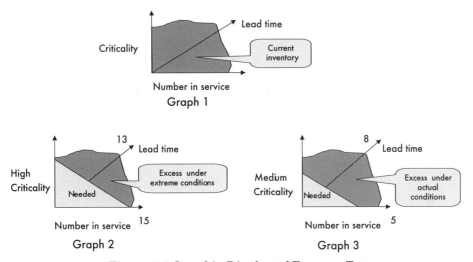

Figure 5.2 Graphic Display of Extreme Test

2 shows the amount of needed inventory determined by the extreme test at a 13-week lead time and assuming 15 units in service (lightly shaded area). The rest of the area (darker area) represents the amount of current inventory for the item that is considered excess. Graph 3 shows that the amount of needed inventory usually decreases, and the excess increases, when more exact values are used for criticality, lead time, and number-in-service.

The gain in excess inventory from Graph 2 to Graph 3 can usually be estimated fairly accurately by multiplying the amount of the Graph 2 excess by a factor of 1.5 for most industries, and by 2.0 for the utility industry. Once the amount of excess for each item is determined, it's an easy matter to sum the results to estimate total gross overstocks.

The extreme test method compares very well to the actual amount of overstocks determined after more precise values are set for the input parameters. At one major U.S. utility, the forecasted amount of overstocks by the extreme test was $60 million; within eighteen months after completing an optimization program, the utility wrote-off $65 million. Regrettably, applying extreme test conditions to determine understocks has not proven to be successful. Using extreme low-end default parameters such as one unit in service, a very short lead time, and low criticality seldom produces MIN/MAX values greater than current values. To determine understocking of an item almost always requires setting actual input values rather than low-ball default values. Consider the following case studies regarding overstocking.

Overstocks are rarely-used key items with an actual balance on hand that exceeds the suggested maximum under extreme test conditions.

Location	Number of Items overstocked	% of total RUKI items	Value of excess above extreme test MAX ($)	Excess as a % of total RUKI value
A	257	6.6	289,235	4.8
B	226	5.2	301,374	3.5
C	39	2.6	27,695	1.3
D	155	7.4	200,846	6.2
E	23	2.5	8,687	0.8
F	73	3.1	123,922	3.6
G	12	6.9	7,716	7.1
H	130	4.6	212,932	4.9
I	347	16.9	490,300	12.0
J	32	7.4	101,721	8.6
K	1	1.3	241	0.9
L	57	7.9	92,737	6.2
M	35	55.6	46,616	41.8
N	666	13.9	1,082,104	12.5
Summary	2,053	7.8	2,986,126	6.7

All of this potential cash flow is simply the result of not replenishing excess stocks as they are used!

This amount is based on the Extreme Test. Overstocks using actual criticality and lead time are probably closer to $4,500,000.

Figure 5.3 Determining Overstocks

CASE STUDY 5-1: AJAX CHEMICAL

The Situation:

AJAX Chemical operated a major complex that was spread over three nearly-adjacent sites in Louisiana. Two years ago, they made a major effort to eliminate obsolete and excess inventory, taking advantage of some good years of profits to write off nearly $2 million of excess inventory. The plant finance director, who was still not satisfied that stocking levels were at optimum level, contacted a consultant who routinely performed a no-cost analysis of potential overstocks. "Hey, as long as I can get a free second opinion, why not?"

The Proposed Solution:

The consultant used an extreme test method to determine potential overstocks, then adjusted the test results using a factor of 1.5, a multiple considered appropriate for the chemical industry.

The Numbers:

Figure 5.3 shows the results of the extreme test. For the 14 storerooms, slightly over 2,000 RUKIs (only 7.8 per cent of the total RUKIs) were found to be overstocked using high criticality and a 13-week default lead time for each item. Total overstocks were determined to be about $3 million, or 6.7 percent of the total RUKI value. Applying the 1.5 multiple resulted in a projection of about $4.5 million of actual overstocks.

The Conclusion:

The finance director and the consultant noticed that most of the overstocks were confined to two storerooms (I and N). They decided to sponsor a program with the consultants to conduct a more thorough analysis limited to those two storerooms.

Industry	Extreme test overstocks (%)	Actual overstocks (%)
Pulp and Paper	12	18 - 24
Transportation	35	50 - 60
Chemical	11	17 - 21
Metals	18	25 - 32
Glass	18	26 - 35
Refining	19	27 - 37
Utility/Fossil	25	35 - 48
Utility/Nuclear	28	38 - 52

Figure 5.4 Typical Results from Extreme Tests

Our experience at over 700 plants has shown that the amount of overstocks projected by the extreme test varies depending on industry (see Figure 5.4). Transportation tends to always be on the high side, whereas chemical and paper companies routinely show the lowest overstock levels. There is a small but not significant difference between nuclear and fossil-fired power stations. It is not surprising that electric utilities carry high levels of excess; they are able to roll the excess into their rate of return approved by state rate commissions for their regulated operations.

CASE STUDY 5-2: AJAX PAPER

The Situation:

AJAX Paper established a team to review some of their stores items last year. During the year, the team reviewed 5,700 items with an inventory value of $742,000; they made numerous adjustments to the current system MIN/MAXs. Later, the plant management asked a consultant to audit the team's work.

The Proposed Solution:

Using Pareto's rule, the consultant determined that the optimum number of items to review (the rarely-used key items) was 6,950 comprising over $15 million of inventory. The extreme test method was used to determine overstocks.

The Numbers:

The results of the extreme test showed an overstock for the 6,950 items of about $3 million. This amount was no surprise. What was a surprise, however, was that only 180 of the 5,700 items reviewed by the team were on the consultant's key item list ($283,000 of the $15 million).

The Conclusion:

The time spent by the team to review the 5,700 items was virtually wasted! Whatever approach they used to select the items, the amount reviewed was less than 2 percent of the Pareto rule amount. Had they planned better, they could have reviewed almost $15 million of inventory instead of only $742,000.

CASE STUDY 5-3: AJAX UTILITY

The Situation:

AJAX Utility operates a large generating station in the midwestern United States consisting of two fossil-fired units and one nuclear unit on the same property. Because all the units were relatively new (the oldest was 10 years old), the plant manager was not overly concerned about what seemed to be an excessive amounts of spares. Besides, the financial people had said they could include the excess in the station rate base, which would guarantee them a good return on the unemployed assets. Recently, however, the plant manager discussed several store items with the maintenance manager and reached this conclusion, "Even if we get paid to keep a lot of the excess, I'm not sure I can defend the large amount we're carrying for some items."

The Proposed Solution:

The plant manager hired a consultant to help analyze a select group of items. An extreme test was used to determine the amount of excess, then suggest a recommended disposal amount for each item.

The Numbers:

Figure 5.5 summarizes the analysis for nine items from four of the station storerooms. Using five years of recent usage history, criticalities set by a team from maintenance, and current material system lead times, the number of units and value of the overstocks for each item was determined using proven stocking codes. Factoring in the usage pattern for each item over the last five years allowed the consultant to determine a recommended disposal amount for each item that was more than adequate to protect each spare to maintenance's desired availability. One to 47 units of each item were recommended for disposal.

The Conclusion:

Of the nine items shown in Figure 5.5, five had a "years supply" exceeding the 40-years remaining life of the station. The station manager agreed to the consultant's recommendations and submitted a request to write-off $1.6 million of inventory in the current fiscal year.

Surely, keeping a lifetime supply is more than enough!

Calculations were performed at 300% replacement cost and *NO* disposal value.

Store	Stock No.	Price ($)	5 yr usage	Excess stock		Years supply	Recommended disposal	
				Units	Value ($)		Units	Value ($)
B	022077	2,700.00	0	30	81,000	226	29	78,300
C	060642	9,838.00	0	5	49,190	38	3	29,514
C	085654	1,810.00	0	12	21,720	90	10	18,100
M	234175	1,438.00	1	14	20,132	43	11	15,818
B	089677	252.19	0	48	12,105	361	47	11,853
R	142778	332.72	13	78	18,152	29	31	10,314
M	676784	354.96	0	25	8,874	188	24	8,519
M	097874	7,811.39	1	4	31,246	12	1	7,811
C	033426	945.47	1	11	10,400	33	8	7,564

▭ Exceeds normal 40-years remaining life of plant

Figure 5.5 The Cost of Excess

5.3   GETTING RID OF EXCESS INVENTORY

Identifying excess inventory can be the easy part. Getting rid of it is the hard part. It's a bit like stepping on a scale and reading you're 20 pounds overweight. Unless a plant expenses spare parts as they are purchased, which is not a common practice, the value of the spares stays on the inventory books until used or disposed. Whether you call it disposal, write-off, investment recovery, or some other term, getting rid of something usually means taking a hit on the bottom-line. We will discuss investment recovery in more detail later in this chapter.

Without question, the best way to work down excess inventory is to use it. Active items will work down quickly because they are used more frequently. We have found that about one-half of the excess rarely-used inventory will work off through usage over a four-to-five year period (see Figure 5.6). The remaining excess rarely-used items, however, may or may not ever work down to desired levels because many of them will never be used during the life of the plant.

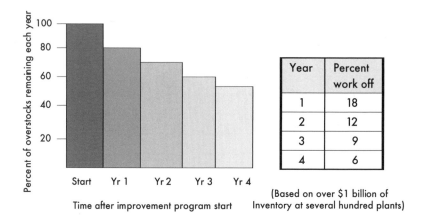

Figure 5.6 Overstocks Will Work Off Over Time

The rate of work off will not decrease to zero after 5 years, but can easily decrease to one or two percent per year by year 10. But remember this. Although the excess is working off, new spares and excesses from overhauls can offset much of the work off decrease.

5.4   STRIVE FOR A BALANCED INVENTORY

Stocking recommendations using decision support codes are always based on less-than-perfect input data. For example, suppliers may say the lead time for a spare part is four weeks, but will they actually deliver by then? The criticality for a spare will always be somewhat judgmental. What is high criticality to one person may be medium criticality to another. Such differences provide all the more reason to perform sensitivity analyses when the input data are spongy.

Maintenance and operating personnel will never completely agree on a consultant's recommendations for every item. Accordingly, we have always considered any current stocking level for an item that is within one unit, higher or lower, of our recommended reorder point to be balanced. The reason is because it usually takes a change of one level of criticality (e.g., from high to medium) or a swing of four-to-six weeks on the lead time, to change the recommended reorder point by one unit.

Most reorder points should be set within one unit of the optimum reorder point.

Difference (optimum − planned OP)	Number of stock units	Percent of total stock units
-6 Or More	8	0.5
-5	1	0.1
-4	3	0.2
-3	6	0.4
-2	29	2.0
-1	116	8.2
0	1,118	78.6
+1	97	6.8
+2	29	2.0
+3	7	0.5
+4	2	0.1
+5	1	0.1
+6 or more	8	0.5

93.6 % of reorder points are in "balanced" zone

Balanced inventory area

Figure 5.7 Achieve A Balanced Inventory

Figure 5.7 shows a profile of the deviation of MINs (planned reorder points) from the optimum MIN for a typical production plant. In this example, 93.6 percent of the system MINs were already at or have been adjusted to be within +/-1 unit of optimum. Don't spend a lot of time trying to fine-tune these items; instead, focus your attention on the items that deviate more than one unit high or low.

5.5 SOME WAYS TO OPTIMIZE THE INVENTORY INVESTMENT

In this section we will discuss several proven techniques for improving the inventory balance. Each method focuses on reducing inventory.

5.5.1. Reducing the Replenishment Lead Time

The inventory investment needed to support production process

---

- Select key items for review by suppliers.
- Prepare supplier request form:
    - Confirm current price and delivery lead time
    - Determine and ask for shorter lead time
    - Quantify if there is price premium for shorter delivery
    - Request comments (some typical replies):
        - Item descriptionis not accurate
        - We no longer manufacture the part
        - Price is subject tochange after 30 days
- Evaluate supplier responses.
- Make changes in material system parameters.

Figure 5.8 Reduce Replenishment Lead Times

availability can be cut in half when vendor deliveries are responsive to plant needs. This finding is based on analyzing inventory require-ments at over 700 production plants worldwide.

In Chapter 3, Section 3.4.2, we showed how stocking level varied as the lead time to replenish increased. Conversely, as the lead time is shortened, less inventory is required, showing the highly leveraged sen-sitivity of stocking level to lead time. Therefore, cutting the replenish-ment lead time can made a significant impact on improving cash flow.

Our experience has shown that lead time data in most material sys-tems is either inaccurate or non-existent, usually due to system lead times being reviewed and updated only as a result of a recent pur-chase. If the part has not been purchased during the last several years (typical for rarely-used spares), here are several potential problems to look for with regard to quality of your lead time data:

- A new material system has been installed which was not populated with any of the previous system's lead time data.
- Vendor(s) could have gone out of business.
- Vendor(s) could have merged with another company and changed names.
- Vendor(s) no longer supplies the part.
- Previous manufacturer(s) no longer makes the part.

- Vendor(s) has relocated (nearer or farther from your facility).
- Part has changed fabrication status from mass-produced to special order or vice-versa.

These problems are not surprising considering that many of the spares have not been purchased for five or ten years, or more. In fact, over one-half of rarely-used spares at most production plants have not been used in the last five years. Therefore, updating lead time data is important. Figure 5.8 summarizes an approach for making these improvements.

It is not necessary or justifiable to update lead time data for all spares in the storeroom. First, focusing only on the key items using the Pareto rule should cut the number of spares being considered to about 10 - 15 percent of the total. Next, the key items can then be tested against the following four criteria; only items meeting these criteria should be considered for further review:

- Items where the current replenishment level is expected to be reached in the next two years.
- Items where a saving of at least $500 is possible.
- Items where at least one unit of inventory reduction is expected.
- Items where a lead time reduction of less than 50 percent would be required.

Normally, only about 40 - 60 percent of the initial items will pass these four hurdles and be candidates to submit to vendors. For each one that is, a specific shorter lead time needs to be identified to the vendor with a request as to whether they can meet it and, if so, whether there is a price premium for doing so. This phase of the program usually requires two separate mailings in order to get a response rate of 80 - 90 percent. Vendors failing to respond are sometimes removed from future purchasing consideration. The following case describes the concept in more detail.

CASE STUDY 5-4: AJAX UTILITY

The Situation: *AJAX Utility is a major electricity-generating company in the eastern United States. AJAX operates eight large and two medium-sized generating stations, all fossil-fuel fired. Recently the company contracted with inventory consultants to help them optimize their large inventory of replacement spare parts. The consultants suggested that they update their vendor lead time information because many of the spares had not been purchase in over 5 years. The supply-chain manager was reluctant to support the lead time update program because it would be too costly, involving too much time by the purchasing department. "Right now we're reviewing over 12,000 key spares to update our stocking levels. Because we use as many as ten vendors on some spares I think it would be prohibitive to solicit feedback from all of them." The consultants proposed a solution.*

The Proposed Solution: *The consultants proposed that the 12,000 key items be tested against the criteria described above (Section 5.5.1) to see how many items cleared all four hurdles. They also proposed that the company provide the needed data so that a form could be sent to each vendor containing certain information and requesting feedback on current pricing, current lead times, and whether or not the vendor could meet a predetermined shorter lead time.*

The Numbers: *A total of 6,267 items cleared all four hurdles, representing items previously procured from 912 vendors. The consultant prepared the forms, which were sent out to each vendor with a cover letter from the buyer. The response rate from the first mailing was 56 percent (about normal); the utility elected not to conduct a second mailing to the non-responding supplies because they wanted to close out the program without delay.*

The Conclusion: *Based on the lead time data updated by the vendors responding, stocking level reductions worth $1,069,200 were made in the material system. Additional findings were: 1) twenty-eight percent of the desired shorter lead times were met by vendors without a price premium, 2) eleven percent of the shorter lead times were met, but a price premium was required, and 3) 186 highly-critical items were identified by suppliers as no longer available for sale. These items were sent to the various generating stations to find alternate sources before existing stocks were depleted. Also, written comments from suppliers were compiled and circulated for consideration and action. These tended to relate to accuracy issues like wrong vendor manufacturing number.*

Two central warehousing approaches:

1. Common items stocked at a single warehouse
2. Common items stocked at the storeroom with the most usage

Other considerations:
1. Storerooms should be close to each other.
2. The master warehouse uses the most.
3. Only the master works with vendors on reorders.
4. Masters will use the vendor's lead time.
5. Satellites will draw from the master warehouse.
6. Satellites will use an internal lead time (e.g., one week)
7. Rarely-used items maybecome active when centralized; they can then be forecasted.
8. Master warehouses should haveadequate spares for all items.
9. Administrative costs of moving stock must be considered.
10. Maintenance prefers to "see"their items.

Figure 5.9 Central Warehousing Considerations

5.5.2. Sharing Common Spares

For storerooms that are in reasonably close proximity to one another, central warehousing of specific parts can often produce a cost savings. The recommended stocking level at the central warehouse will be lower than if the same stock was held at each storeroom, due to the shorter delivery lead time from the central storeroom to the outlying facilities compared to that of the supplier. The benefits of common stocking can usually be obtained without the need to locate common spares physically at the central site. Instead, a Master/Satellite concept is often employed, in which the master warehouse for any one part is determined by a combination of up to three possible factors, in any order of priority:

- Warehouse supporting the highest usage rate.
- Warehouse with the largest quantity on hand.
- Warehouse with the closest proximity to the preferred supplier.

All other sites become satellite warehouses.

As discussed in the previous section, lead time plays a key role in setting the stocking levels for spares. It also plays a key role in optimizing the stocking of inventory in a shared-spares environment. In fact, it is the shorter lead time (delivery time) from the master warehouse to the satellite warehouses, compared to each procuring from the vendor, that accounts for the benefits obtained from central stocking. Within a storeroom, some items have more than enough stock and other have too little. Although overall inventory investment may be appropriate, opportunities exist to improve service and availability by adjusting by-item imbalances. Increased centralized stock planning and management concepts can help by expanding existing informal shared inventory concepts into a more formal centralized program for selected items.

Two approaches are frequently used for central warehousing, as shown in Figure 5.9: 1) commonly stocked items are moved to a single central warehouse and distributed to the other sites from there, and 2) common items are stocked at the storeroom with the highest usage, although one unit for highly critical spares may be kept at each of the other sites. There is an additional benefit to centralized stocking of common spares. Frequently, an item that is rarely used at each site will have sufficient usage to qualify as an active item when centralized and, therefore, can be forecasted.

The choice of which approach to use for centralizing inventory depends on a number of factors including warehouse space, actual delivery time between plant sites, and the administrative cost of moving stock and managing the program. An important intangible issue frequently comes into play when a centralized stocking program is considered. Maintenance personnel like to be able "see and touch" their spares inventory. When the spares are off site they tend to worry. This tends to favor the master/satellite concept where items are stocked at the site that has the most usage. Other considerations are summarized in Figure 5.9. Let's look at a case study that covers central stocking.

CASE STUDY 5-5: AJAX ELECTRONICS

The Situation:

AJAX Electronics manufactures a variety of circuit boards and other electrical components for sale to the general industry. The company operates out of three assembly plants in Texas, Louisiana, and Arkansas; many commonly-supplied assemblies are manufactured at each plant. Up to now, reorders of production spare parts to keep plant machinery running were handled at each plant, as was stocking.

The supply-chain manager, who just returned from a lecture on the benefits of central stocking, told the general manager, "I think we should look into central stocking of common spares. I especially liked the master/satellite concept where the plant with the most usage of an item gets to be the master warehouse and all the others draw from them. I think that would do a lot to put the fears of maintenance to rest and save us a lot on carrying costs".

The Proposed Solution:

The general manager approved hiring the lecturer to present an analysis of the benefits of central stocking. It was agreed that the presentation would focus on a single spare selected by the company. Figures 5.10 and 5.11 show the highlights of the consultant's presentation.

The Numbers:

Figure 5.10 shows the failure rates and current stocking levels for the $10,000 spare stocked at each of the

All spare are:
• Considered highly critical   • Cost $10,000

Part number	Annual failures	MIN	MAX	Lead time (Weeks)	Average inventory ($)
ABC-Store1	2	3	4	20	34,954
ABC-Store2	3	4	4	20	32,895
ABC-Store3	1	2	3	20	26,831
Total			11		94,680

Figure 5.10 Shared Spares Before Central Stocking

three sites before central stocking (part ABC-Store1 was stocked at the Texas plant; ABC-Store2 at the Arkansas plant; and ABC-Store3 at the Louisiana plant). Using the parameters shown, the total average inventory required to support the desired availability was $94,680. The average inventory is the amount expected to be on the storeroom shelf at any point in time. Notice that the Arkansas plant had the highest usage rate of 3 failures.

Using the master/satellite concept stocking approach, the storeroom with the highest annual usage (Arkansas) was designated as the master warehouse and the other two are satellites (Figure 5.11). A total stocking decrease of three units was achieved, from 11 to 8 units. This decrease created an annual savings of $28,206, nearly 30 percent. In this case, the lead time to replenish the satellites is cut from 20 weeks to one week, the internal time considered reasonable to transfer the item from the master to the satellite warehouses.

Satellite Items are Transferred Internally

Part number	Annual failures	MIN	MAX	Lead time (Weeks)	Average Inventory($)
Satellite					
ABC-Store1	-	0	1	1	9,751
ABC-Store3	-	0	1	1	9,843
Master					
ABC-Store2	6	5	6	20	46,880
Total			8		66,474

Figure 5.11 Shared Spares After Central Stocking

The Conclusion:

When the overhead and handlings cost were factored into the analysis, the net benefit for the item dropped to $17,301. Based on the presentation, the general manager approved a program to extend the analysis to all common spares with an average unit cost of $1,000 or more.

The decision to limit central stocking in the above case to only items with a cost of at least $1,000 was justified. At a common stocking analysis for one client,

2,434 rarely-used items were used at multiple sites, but only 164 (7 percent) of the items qualified for master/satellite stocking after overhead and handling costs were considered. However, for these items, a 36 percent inventory reduction was achieved, with over 80 percent of the cash flow benefits coming within one year of implementing the central stocking arrangement. Had more items been centrally stocked, the total benefit would have been reduced. It is important to know which items to stock centrally and which not to stock! Let's look at another case study.

 CASE STUDY 5-6: AJAX UTILITY

The Situation:

AJAX Utility is a southeastern generator of electricity covering four states from Florida to Virginia. Besides generating electricity, it owns one of the largest transmission and distribution (T&D) networks in the United States. Because it operates a large number of sub-stations, it has traditionally stocked transformer spares at several regional warehouses in each state. The supply-chain manager, who was becoming concerned about the reorder points for several high-priced spares at the eight regional warehouses in Georgia, was pretty sure stock levels could be cut without sacrificing availability.

The Proposed Solution:

A master/satellite concept was considered to replace the current practice of allowing each regional warehouse to set its own stocking levels and replenish at each site.

The Numbers:

Current reorder points and replenishment lead times for the transformer regulator are shown in Figure 5.12 for each of the eight locations. Because Location 6 had the highest usage for the regulator, it was chosen as the Master warehouse and was assigned responsibility for dealing with the vendor. Each of the other locations would replenish from the master at a one-week lead time and adjust its reorder point accordingly. As a result, Location 6 increased its reorder point from 3 to 8 units, while all the other locations cut their reorder point from 1 or 2 units to zero.

5.5.3 Disposal of Gross Excess Inventory Investment Recovery

Identifying overstocks is more than finding items with a quantity on hand greater than the current maximum. Actually, the current maximum may not accurately reflect the most you want to have. Typically, initial minimums and maximums have been determined by vendor recommendations or suggestions from maintenance personnel. More often than not, this quantity was too high. Over time, a few maximums may have been reduced, but if the storeroom ever was caught short, that item's maximum was raised to a quantity which more than accommodated the "worst ever" condition.

Companies not using an inventory decision support tool usually have valid concerns about the correctness of their maximums and do not have organized information regarding the criticality, lead time, and number in service for each item. Some people actually view the absence of any one element of this information as a convenient excuse for doing nothing. As we showed earlier in this chapter, the extreme test method for identifying overstocks does not require prior knowledge of any of this information and requires no assistance from plant personnel. If the current quantity on hand for an item exceeds the maximum calculated by the extreme test, then a logical conclusion is that the on-hand quantity is very likely to be too high. To test this hypothesis, we applied the extreme test to 125,000 rarely-used items with $500 million of inventory value. These items were later evaluated after clients set more accurate parameters and reviewed the results. Here's what we found:

- The extreme test recommended 60 percent more inventory than when actual values were used. However, nearly 40 percent of the items had an actual on-hand quantity greater than the maximum suggested by the Extreme Test.

- By using actual values for the three inputs (criticality, lead time, and number-in-service), clients confirmed that: 1) 99 percent of the items exceeding the extreme test actually were overstocked, 2) 43 percent of the total inventory investment was tied up in overstocks above and beyond levels required to support operations properly, and 3) anticipated usage eliminated only about one-half of the overstocks during the next four years.

- Some items **not** failing the extreme test were also found to be overstocked when evaluated with actual values for criticality, lead time, and number-in-service. However, most all of these were overstocked to such a minor degree that normal usage would eliminate virtually all of their overstocks during the next four years.

Only 3 percent of total overstocks that would not be eliminated during the next four years went undetected by the extreme test.

The part Is a transformer regulator costing $3,200

Location	Current LT (Weeks)	Current OP	New LT (Weeks)	New OP	Central store (Yes/No)	OP change
1	13	1	1	0	No	-1
2	13	2	1	0	No	-2
3	13	2	1	0	No	-2
4	13	2	1	0	No	-2
5	13	1	1	0	No	-1
6	13	3	13	8	Yes	+5
7	13	1	1	0	No	-1
8	13	1	1	0	No	-1
Total	-	13	-	8	-	-5

Central warehouse dealing with vendor 13-week lead time

Figure 5.12 Central Warehousing

We have, therefore, determined that every production storeroom has a substantial overstocking of spares, and the extreme test can accurately identify which items are surplus. Yet what can we do about it? The first thing we need to do is recognize where much of the excess actually comes from, as shown in Figure 5.13. Then we must deter-

Excess inventory accumulates from these sources:

- Buying too many initial spares
- Replenishing before necessary
- Buying too much for planned overhauls
- Misidentifying parts or not using them for their intended purpose
- Never removing obsolete items from inventory
- Inheriting items from another plant

Figure 5.13 Sources of Excess Inventory

mine how to get rid of most of it. Here is where investment recovery comes into play.

5.5.4 The Investment Recovery Matrix

Although investment recovery is the way to address excess inventory, most companies limit their efforts with the excuse, "I don't have the budget for investment recovery write-off." This failure to act in the overall best interest of the company burdens operations with substantial overstocks and costs that can and should be avoided. Contributing in turn to the costly inaction is the decision whether or not to **sell** any of the excess. That decision can be a complex, multifunctional question best handled by a financial model—the Investment Recovery Matrix (IRIX).

This comprehensive econometric model considers all of the parameters indicated in Figure 5.14; it recommends the quantity to dispose, if any, that produces the greatest present value of the resulting company-wide net cash flows. IRIX generally takes the combined inputs from several sources in a company to provide the values needed to calculate the matrix results. The current book value, reorder point, and quantity on-hand can usually be obtained from the material system. Most of the other parameters can be provided by the financial or operating personnel, leaving the replenishment cost per unit and the disposal value per unit as the two major variables in the matrix. Appen-

Prudent investment recovery decisions need to consider these parameters:

- Current book value of each unit
- Actual quantity on hand
- Current replenishment cost per unit
- Inflation rate of replenishment cost
- Disposal value per unit
- Cost of capital rate
- Variable carrying cost rate
- Corporate income tax rate
- Years to be considered in the analysis (e.g., 15 years)
- Reorder point
- Economic order quantity (EOQ)
- Cost of writing a purchase order
- Expected annual usage

Figure 5.14 Disposal of Gross Excess Inventory

dix 5-1 to this chapter explains each of the IRIX parameters in more detail.

IRIX uses eight user-defined multiples of current book value per unit for the Replenishment Cost and eight other user-defined multiples of current book value for the Disposal Value. For each of the sixty-four possible combinations, the model evaluates the present value of the net cash flow for every possible action ranging from doing nothing to disposing of all excess stock.

Figure 5.15 shows a sample IRIX output for a part with a current book value of $1,000. The maximum planned stock level was 3 (MIN of 2 plus and EOQ of 1). With a quantity-on-hand of 14, there are 11 excess units. Therefore, for each of the sixty-four combinations, twelve alternative actions are evaluated ranging from disposing of 0 to disposing of 11. The most economic disposal quantity is printed in the matrix for each of the sixty-four combinations.

Suppose the replacement cost is $1,100 and the disposal value is $150. After running the alternative actions for that combination, you determine that it would be best to dispose of 7 units. That decision is reflected in Figure 5.15. Similarly, if the replacement cost is $1,500 and the disposal value is $150, there is no disposal amount. At a re-

Replacement cost (Dollars)

Disposal value (Dollars)	750	1,000	1,100	1,200	1,300	1,400	1,500	2,000
0	9	6	5	0	0	0	0	0
50	9	7	6	5	0	0	0	0
100	10	7	6	5	4	0	0	0
150	11	8	7	6	5	4	0	0
250	11	9	8	7	7	6	5	0
500	11	11	11	10	9	8	8	4
750	11	11	11	11	11	10	10	7
1,000	11	11	11	11	11	11	11	9

Current book value of item = $1,000

Figure 5.15 Sample Investment Recovery Matrix

placement cost or $750 and no disposal value, 9 is the disposal amount.

To implement the IRIX recommendations, look to see if at least some stock should be disposed over a range of likely current replacement cost and disposal value combinations. If the matrix has nonzero's in the likely combinations, contact the vendor to obtain the current replacement cost, then highlight the column with the closest replacement cost. If any of the "likely to achieve" disposal values in the highlighted column recommend selling at least some of the excess, attempt to sell the item. When an offer is received for an item, decide whether or not to continue shopping for a higher offer or else sell the quantity suggested by the model. Selling that quantity, even if for only five or ten percent of current book value, has been determined by the econometric model to maximize the present value of the overall net cash flows, in and out, to the company. Figure 5.16 lists some of the more likely venues for disposing of excess inventory, in order of decreasing return on the disposal.

- Use excess in near-term scheduled maintenance.
- Return to vendor.
- Transfer excess to a sibling facility in the same company.
- Sell to another local company.
- Sell to another company in your industry (yes, even competitors!).
- Sell to scrap dealer.
- Deposit in dumpster.

Figure 5.16 Some venues for Investment Recovery of Excess Inventory

Replacement cost as a percent of book value)

(Quantities shown are most economic to dispose)

		110 %	125 %	150 %	175 %
Disposal value as a percent of book value	0 %	878	800	613	556
	5 %	900	820	622	570
	10 %	918	837	673	598
	20 %	951	892	787	611

Replacement cost = 150 %   Disposal value = 5 % 15-year analysis

Figure 5.17 Number of Items To Dispose

In our large, multi-client database containing millions of items, over forty percent of the total inventory is tied up in overstocks. Using typical input parameters, conservative assumptions that for all items, the current replacement cost is only 120 percent of book value, and the best achievable disposal value is only 5 percent of book (scrap) value the model determined that the best decision is to sell almost one-half of the overstocks and keep the rest until eventually used by operations in future years. The following case study discusses this further.

CASE STUDY 5-7:   AJAX NUCLEAR

The Situation:

AJAX Nuclear operates two 980-megawatt nuclear reactors at a site in central Illinois. A recent inventory initiative determined that a substantial amount of slow-moving spare parts were excess. A pool of $5 million had been set aside for the disposal and write-off of surplus, but the company's board of directors wanted to see a list of the items planned for disposal before approving the use of the pool. A consultant with experience in investment recovery was hired to help prepare the list.

The Proposed Solution:

The consultant proposed using an econometric model to determine which items should be considered for disposal.

The Numbers:

Using the IRIX model described above, the consultant and the company managers agreed on the following economic parameters to simplify the disposal decision: 1) a replacement cost for each item of 150 percent of current book value, 2) a disposal value of 5 percent of book, 3) a 4-percent inflation rate, and 4) a fifteen-year analysis period. All other parameter values were provided by either the supply-chain manager or the financial manager.

Figures 5.17, 5.18 and 5.19 show the results of the investment recovery analysis. The highlighted block in each figure marks the intersection of 150 percent of book value with a 5 percent disposal value. Under these parameters, 622 stock items were determined to be candidates for disposal; these items had a current book value of $2.884 million. The economic cash flow savings, mostly from tax credits from disposing of items below book value, were determined to be $4.151 million.

The Conclusion:

The board of directors approved the use of the pool to dispose of the 622 items. It further stipulated that any funds obtained from disposal in excess of the 5 percent assumed could be retained by the station for maintenance and operating programs.

5.5.5 Consignment Of Selected Parts

Management is always looking for ways to cut costs. Sometimes they resort to sleight-of-hand maneuvers such as laying employees off just to rehire them as contract workers, in order to cut head count. Although the lower head count may look good in the annual report, the cost savings from this tactic, if any, may not be significant.

The same maneuver is frequently used to move inventory assets off the books, yet still have a large degree of control over the material. This is called consignment. Consignment inventory is defined as inventory that is still controlled by the customer, but is owned by a supplier. It can work two ways: 1) the company can transfer ownership of some of its existing stock to a supplier, who will retain it and only charge the company when stock is drawn from the storeroom, or 2) the supplier can place some of its inventory in the company's storeroom and only charge when used. This last case still obligates the company to store and handle the inventory at some cost. Exactly where the inventory is located can also vary, either on-site within the plant, adjacent to the plant, or at a remote location.

The type of spares to be consigned is also a consideration. Consignors prefer to handle active items that turn over frequently rather than slow-moving stock. Typically, the consignor only gets paid when

Replacement cost as a percent of book value

(Amounts shown are most economic to dispose) $000

		110 %	125 %	150 %	175 %
Disposal value as a percent of book value	0 %	$4527	$4066	$2843	$2490
	5 %	$4663	$4174	$2884	$2587
	10 %	$4756	$4287	$3139	$2693
	20 %	$5026	$4619	$3989	$2817

Replacement cost = 150 %   Disposal value = 5 %   15-year analysis

Figure 5.18 Value of Disposed Inventory

Replacement cost as a percent of book value
(Amounts shown are economic savings from disposal) $000

		110 %	125 %	150 %	175 %
	0 %	$4961	$4498	$3970	$3659
	5 %	$5238	$4749	$4150	$3820
	10 %	$5521	$5004	$4334	$3987
	20 %	$6106	$5545	$4791	$4335

Disposal value as a percent of book value

Replacement cost = 150 %   Disposal value = 5 % 15-year analysis

Figure 5.19 Economic Savings from Disposal

the item is used in the production process or in a maintenance procedure, although other payment plans are occasionally used. If the stock
item is critical to maintaining production, it may be necessary to have
the consignor reserve a certain quantity of the item to insure its availability when needed. We will discuss a concept called "Certified Availability" in Chapter 9 that does just that. Otherwise, the consignor may
not always have enough material in stock to meet both your needs and
that of its other customers.

Regardless of who owns the stock, or where it is physically stored,
the concept of consignment should be invisible to order processors and
warehouse personnel; it should not cause disruptions to the normal
shipment and receiving processes in place for non-consigned items.
Otherwise, it may be time to restructure the current supply-chain
process to incorporate consignment. With many consignors available,
it may provide a significant financial advantage to use one.

APPENDIX 5-1

Explanation of IRIX Terms

Variable	Effect	Explanation
Current book value	None or Direct	The base value from which many of the calculations are made. Changing this value, while keeping all other inputs the same, usually has no effect on the quantity to be disposed. Higher values do increase the absolute value of savings, but not the relative value.
Quantity-on-hand	Direct	Higher values increase the likelihood of overstocks and the potential quantity to be disposed.
Current Replenishment Cost	Inverse	Higher values increase the cost at which future replenishments will be made and may cause fewer to be disposed.
Disposal Value	Direct	Higher values increase the immediate cash flow from disposals, causing a trend for more to be disposed.
Annual Usage	Inverse	Higher values mean the overstocks will be eliminated by usage at a faster rate, causing a trend for fewer to be disposed.
Minimum/Maximum	Inverse	Higher Maximum values (Minimum plus EOQ) reduce the amount of overstock, causing a trend for fewer to be disposed.
Inflation Rate	Inverse	Higher values cause both future replenishments and future purchase orders to cost more; may cause fewer units to be disposed.
Cost of Capital	Direct	Higher values increase the benefit of receiving cash now from the disposal value plus the income tax effect of write-offs, resulting in more to be disposed. The cost of capital is the value by which the present value of future cash flows, in and out, is calculated. Higher values reduce the present value of the cost of future replenishments; this also may cause more to be disposed.
Variable Carry Cost	Direct	Higher values make it more expensive to hold inventory, causing a trend for more to be disposed.

continued on next page

Explanation of IRIX Terms (continued from previous page)

Variable	Effect	Explanation
Purchase Order Cost	Inverse	Higher values make future replenishments more expensive and may cause fewer to be disposed.
Income Tax Rate	Direct or Inverse	Higher values increase the cash received from or write-offs (disposing for less than book value) and tend to result in more to be disposed. However, in the unlikely event that the disposal value is greater than the book value, taxes must be paid on the gain, causing a trend for fewer to be disposed.
Minimum Difference	Inverse	Unless the present value of the net cash flow resulting from all of the above factors is at least the minimum difference amount better than disposing of none of the overstock, IRIX will recommend doing nothing. This eliminates recommendations for disposal, which, although better than doing nothing, are of a benefit too small to justify the effort.
Number of Years	Inverse	Higher values increase the likelihood of replenishments (at the current replenishment cost plus the compounded inflation effect) having to be made during the period being evaluated, causing a trend for fewer to be disposed.
General Note		All overstocks on-hand not expected to be used by the end of the evaluation period are to be disposed. Because IRIX produces substantially more prudent and defensible result than using a fixed-number-of-years' supply guideline, we suggest using a value between ten and twenty years rather than some arbitrarily shorter period.

CHAPTER 6

AVOIDING UNNECESSARY PURCHASES

6.1 WHAT THE READER SHOULD LEARN FROM THIS CHAPTER:

- Twelve ways to avoid unnecessary purchases
- A procedure for buying spares when the reorder point is too high
- How unnecessary purchases vary from company-to-company
- Consequences from cycle count errors

6.2 TWELVE ACTIONS THAT CAUSE UNNECESSARY PURCHASES

It didn't take us very long to come up with the following twelve examples of actions that lead to making unnecessary purchases:

- Not taking advantage of the benefits of consigning
- Not taking price/quantity discounts when available
- Failing to partner with vendors
- Buying too early because the MIN is set too high
- Not ordering the economic order quantity (EOQ)
- Overbuying initial spares
- Ordering to inaccurate (longer) lead times
- Not achieving the minimum lead time possible
- Commingling overhaul parts with safety stock
- Not considering the incremental MAX
- Failing to weed out duplication
- Buying too many spares because of cycle count errors

Some of these examples have been discussed previously whereas some will be discussed in later chapters. All will be covered in this chapter as well.

6.2.1 Not Taking Advantage of the Benefits of Consignment

We first discussed consignment in Chapter 5 as one way to reduce storeroom inventory, while taking some of the inventory asset off the balance sheet. In consignment, a vendor takes over the responsibility for carrying and replenishing selected stock items. The customer pays for the item only when drawn from the storeroom or ordered from the consignor's warehouse. If managed effectively, the consignor can take advantage of quantity discounts and other techniques such as forecasting to keep the average cost of the consigned stock to a minimum.

In Chapter 9, we will document a case study on the benefits of consigning a portion of the inventory.

6.2.2 Not Taking Price/Quantity Discounts When Available

In Chapter 4, Section 4.4, we showed a case study covering price discounts for ordering larger quantities of an item. Unit price discounts for larger purchase quantities have been around almost forever. But sometimes buyers fail to ask for them, thereby forfeiting the opportunity to save money. As the case study shows, buying more than you may need to get a price discount is a trade-off against the higher carrying cost of keeping too much inventory too long. Although it can be time consuming to calculate whether or not there are benefits of accepting a price/quantity discount, preparing those calculations is what the purchasing buyer is paid to do.

6.2.3 Failing to Partner With Vendors

Several case studies in Chapter 9 discuss the benefit of partnering with your distributor or original equipment manufacturer (OEM). Frequently, both parties will offer assured stocking programs that allow the customer to carry less of certain items because the vendor will supply the items, when needed, within a prescribed lead time that is usually shorter than the normal lead time. Sometimes the concept of partnering with a supplier goes against the grain of purchasing managers because they sense a loss of the traditional adversarial relationship of hard-nosed negotiation common over the years. But times have changed in the modern supply-chain environment, and partnering with vendors is a common practice today. Failing to do so may lose an opportunity for significant savings on continuing purchases.

6.2.4 Buying Too Early Because The MIN Is Set Too High

We've learned during our years of consulting that having the best material management system in the world is worthless if it's loaded with irrelevant reorder points and reorder quantities. Buying too soon, or buying too much, can be just as bad financially as buying too late and having to accrue the expense of expediting the order to avoid production disruptions. A lot is heard about late deliveries, many of which are due to late placement of the purchase order, but you seldom hear much about the material that is ordered weeks or months early because the reorder point is set too high. Yet these early orders can be the silent killer, sitting on the storeroom shelf, gathering dust, and eating up carrying costs. Even the average homemaker knows the benefits of buying at the proper time, not too early and not too late. If you don't think so, check your pantry at home in June. You will probably not find cranberry sauce, but chances are it will be there the week before Thanksgiving.

About eight years ago we conceived the concept of the extreme test to identify overstocks without requiring support from clients (see Chapter 5 for more on the extreme test). A few years later, we tweaked the test to provide a way to predict items that are being purchased too soon because of inflated MINs. Figure 6.1 shows the procedure used to spot items that should have their replenishment deferred. Once identified, it is necessary for plant personnel to verify the criticality, lead time, and set size, then re-analyze the item to calculate a more precise MIN/MAX.

If the new MIN is lower than the old MIN, the purchase can be deferred. It may also be possible to cancel previously issued purchase orders if the lead times are long. To avoid overburdening buyers, a cutoff based on the price of the item can be set, such as $500, below which no action is required even though there may be benefits.

- Review recent history to project when spares will work down to the current reorder point (ROP).
- Using extreme parameters, determine if the current reorder point is likely to be set too high.
- Use most-likely parameters to suggest a new lower reorder point.
- Review all items where the new lower ROP is below the current ROP to see if any are now in the procurement process.
- Defer purchases on all items where management believes the new lower ROP is valid.
- When possible, cancel already existing purchases orders.
- Repeat the procedure routinely as other items enter the procurement process.

Figure 6.1 Procedure For Determining Unnecessary Purchases

Company	Plants	Industry	Total inventory reviewed ($000)	Estimated excess $000)	Number of early buys	Value of early buys ($000)
A	6	CHEMICAL	53,695	9,540	592	1,769
B	2	CHEMICAL	10,752	2,720	892	1,099
C	1	UTILITY	4,291	1,980	273	349
D	1	CHEMICAL	11,652	2,097	84	179
E	1	PAPER	15,431	3,395	1,025	1,324
F	1	UTILITY	12,716	3,680	393	1,305
G	2	CHEMICAL	23,334	6,002	1,612	3,700
H	1	UTILITY	1, 843	512	11	9
I	1	UTILITY	12,007	2,641	93	227
J	3	CHEMICAL	38,400	4,941	4,192	2,472
K	4	PAPER	44,555	7,251	3,216	3,721
L	1	REFINING	5,507	881	145	330

Figure 6.2 Unnecessary Purchases can vary considerably

Figure 6.2 compiles the amount of unnecessary purchases found during previous studies of twelve companies in various industries. The value of suspect purchases varies considerably, and seems to have no correlation with the amount of excess identified through the extreme test. Most likely, the controlling factor is the percentage of the RUKIs that are currently in the purchasing queue, which can vary widely from plant to plant. Consider the following case studies for AJAX Paper and AJAX Metals.

 CASE STUDY 6-1: AJAX PAPER

The Situation:
AJAX Paper is a producer of linerboard located in the southeastern United States. Like all paper companies, it goes through swings in its business cycle from boom to bust. Business is currently off due to a recession. As a result, cost cutting at the plant has become a top priority. "We still have to continue to produce linerboard," said the supply-chain manager, "but we now look at every purchase carefully before buying anything. Let me show you a recent analysis we just finished on some spares replenishments."

The Proposed Solution:
Consultants hired by AJAX recently completed an analysis of the inventory (see Figure 6.3). They suggested reviewing 121 suspect purchases.

The Numbers:
From the consultant's analysis, 782 stock items were determined to be excess to meet the needs of production. All had current reorder points set higher than necessary. When all 782 items were treated as highly critical, 121 were forecasted to be ready to enter the purchasing process within the next 18 months.

The Conclusion:
Upon review of each of the 121 items ($121,000 worth), the replenishment of 54 items was deferred improving cash flow by $61,400.

AJAX Paper had the following situation:

- A $10,745,000 storeroom inventory consisting of 33,000 items.

- 97% of the inventory was slow-moving (< 12 issues/year).

- 82% of the items had no demand over the last 24 months.

- 782 items were determined to be overstocked by $2,520,000.

- Using high criticality (99.9% availability) and the purchasing system replenishment lead times, 121 items had a recommended reorder point lower than the current system reorder point.

- All 121 items ($121,000 worth) already have hit, or were expected to hit the reorder point within the next 18 months.

Figure 6.3 How AJAX Paper Avoided Unnecessary Purchases

CASE STUDY 6-2: AJAX METALS

The Situation:

AJAX Metals was riding an upswing in their business over the last year. As a result, production increases for their stainless steel and high-alloy finishing lines were scheduled for the next two years. The comptroller wanted to get a good projection of cash requirements for the next two years in order to ensure ample funds were available to cover work-in-progress requirements. She asked the supply-chain manager for his projection on spare parts needs over the next two years for the company's eight storerooms.

The Proposed Solution:

The supply-chain manager provided a list of requirements based on a recent analysis from a consultant.

The Numbers:

Two lists of projected purchases were provided: 1) those items up for replenishment now, and 2) those likely to need buying over the next 18 months. The two lists totaled nearly $4,500,000 (see Figure 6.4).

The Conclusion:

The comptroller and supply-chain manager agreed that further review of the projected purchases lists should lead to deferment of $840,000, a little under 20 percent of the original budgeted amount, beyond the two-year window. Therefore, the budget for spares was set at $3,560,000.

These eight storerooms were planning *Now* and *Soon* purchases (over the next 18 months) and their current reorder points were set too high

Location	Up for purchasing NOW ($)	Up for purchase SOON (18 months) ($)	Total purchases ($)
41	256,000	632,000	888,000
16	183,000	657,000	841,000
03	104,000	228,000	332,000
18	149,000	277,000	425,000
06	159,000	321,000	480,000
12	102,000	342,000	444,000
36	154,000	280,000	434,000
42	167,000	448,000	615,000
Total	1,274,000	3,185,000	4,459,000

Figure 6.4 Avoiding Unnecessary Purchases

6.2.5   Not Ordering The Economic Order Quantity (EOQ)

The consequences of not ordering the economic order quantity can be severe, as shown in Figure 4.6. Buying less than the EOQ increases the order cost, whereas buying more increases the carrying cost. Even when the EOQ is the amount ordered, you have different options regarding when to place the order. Timing can impact overall costs.

Over the last twenty years we have been amazed at the EOQ formula's lack of use by most purchasing departments at client plants. Much of the reason is due to the MIN/MAXs having been set originally by guesses from production personnel who are reluctant to change the values unless they have led to shortages (backorders) in the past. Seldom, did the values get reduced.

6.2.6 Over-Buying Initial Spares

Purchasing initial spares is tricky, especially for slow-moving items, because there is no usage history upon which to base the order. But software tools are now available that can help with the decision on buying new spares. The following two case studies for fossil-fired and nuclear utilities show the magnitude of the potential benefits.

CASE STUDY 6-3: AJAX UTILITY

The Situation:

AJAX Utility is an eastern U.S. utility; it operates generation and transmission facilities in two states. Because of the buildup in their summer peak demand from air conditioning, and in their winter peak from the installation of electric heating in more new home, they decided to add a new combined cycle peaking unit at one of their existing sites. As part of the capital appropriation, slightly over $2,000,000 was allocated for the purchase of new spares based entirely on the recommended stocking levels from the equipment manufacturer for the new unit. Because the project was experiencing cost over-runs, the chief financial officer requested that orders for spares be cut as much as possible.

The Proposed Solution:

The company hired an outside consultant to help them evaluate the new spares situation, using a decision support tool designed to accommodate new spares.

The Numbers:

A team was put together consisting of the consultant and a previous maintenance superintendent, now retired, who was familiar with other units at the site similar to the new peaking unit. Each of the 576 planned new spares was reviewed for criticality, lead time to replenish, and whether it had to be purchased in a certain set size. The consultant's recommended MIN/MAXs for the new spares were then reviewed by the maintenance superintendent who, using personal experience, cut the recommended amount for about 15 percent of the items (see Fig. 6.5). A revised buy list was then set to the CFO.

The revised purchase list cut the original total order size from $2,078,600 to $1,385,300, a near-term cash flow saving of $693,300.

The Conclusion:

The management agreed to use the same team approach when ordering new spares for future capital projects.

A station of AJAX Utility had the following situation:

- They planned to add a new 500 MW peaking unit.

- The supplier recommended they buy 576 separate spares for $2,078,600.

- A team was created to assess the need for the recommended spares.

- Criticalities and mean-time-before-failure (MTBF) data were collected, then analyzed by a consultant to set recommended MAX levels.

- Recommended MAX levels were reviewed by the team; adjustments, up or down, were made as appropriate.

- The actual amount of spares purchased was reduced to $1,385,300, a saving Of $693,300.

Figure 6.5 How AJAX Utility Avoided Unnecessary Purchases

CASE STUDY 6-4: AJAX NUCLEAR

The Situation:
Two consultants were visiting AJAX Nuclear to discuss some work they were doing on another project when the station maintenance manager approached them with this question, "I hear you folks have some high-powered tool that can set stock levels for existing spares. Is it any good for new spares?" The manager showed them a list of ten nuclear valve spares that had to be purchased soon, a potential $800,000 order.

The Proposed Solution:
The consultants agreed to review the items on the spot, with the maintenance manager providing the criticality, mean-time-between-failure (MTBF) estimates, and lead time inputs. Each item was then processed through the decision support tool and a recommended purchase amount determined.

The Numbers:
Figure 6.6 shows the comparison between the vendor's recommended purchase amount and the consultant's recommendation.

The Conclusion:
The maintenance manager agreed to accept the consultant's recommendation, saving over $200,000 on the purchase order. Not a bad effort for a couple of hours' work!

Ten vendor recommended valve spares were reviewed and all were found to be excessive. Purchasing less did not increase risk.

Total vendor recommended order = $800,000   Total savings = $207,845

Valve	Price ($)	Recommended by vendor	Amount purchased	Savings ($)
1	11,558	4	2	23,116
2	8,433	8	4	33,732
3	8,378	4	3	8,378
4	38,033	4	3	38,033
5	32,686	4	3	32,686
6	15,516	4	2	31,032
7	4,806	4	2	9,612
8	5,278	4	2	10,556
9	6,923	4	2	13,846
10	3,427	4	2	6,854

Figure 6.6 Save Money When Buying New Spares

6.2.7 Ordering To Inaccurate (Longer) Lead Times

In Chapter 3, we discussed the "lead time bias" the tendency to overstock items with short lead times and understock items with long lead times. Because the majority of items in a production storeroom are slow-moving, lead times are likely to be inaccurate even if they are updated after each purchase. The competitive pressure experienced by suppliers over the years has forced them to be more responsive to their customers, meaning offering shorter lead times. Our work has shown that at least one-half of the inventory in production storerooms could be eliminated if suppliers could deliver replenishments every time in two weeks or less. To see why, consider the following case study.

CASE STUDY 6-5: AJAX CHEMICAL

The Situation:

AJAX Chemical was undergoing a process upgrade to add capacity to their paint additives line at the Houston facility. An extensive list of new spares purchases had been prepared and was the focus of a meeting between the plant manager and the supply-chain manager. "A lot of these items look like off-the-shelf items to me", said the plant manager, "so why does it take so long to get them? I bet we could cut a lot of inventory if we could get this stuff a couple of weeks quicker."

The Proposed Solution:

The supply-chain manager asked an inventory consultant to determine the impact on stocking levels if the lead time could be cut across-the-board by 25 percent.

The Numbers:

Table 6.7 shows that, if the current lead times shown were cut by 25 percent, the current MIN could be cut by the amount shown, saving over $31,000 for the nine items. Notice that for Part 037899, a 25-percent-lower lead time would make the item a candidate to not stock at all (MIN = -1).

The Conclusion:

Prior to placing the purchase orders for the spares, each vendor was asked to supply the items two weeks sooner than the quoted lead time. Most were able to comply and the quantity actually purchased was set at the suggested MIN in Figure 6.7 plus the EOQ.

6.2.8 Not Achieving The Minimum Lead Time Possible

How do you know if the lead time provided by a vendor or distributor is the best possible lead time? The answer is: You can't know unless you ask. We showed in Chapter 3 that stock levels increase as the lead time gets longer. Then, in Chapter 5 we described a process for determining in advance what lead time reduction to ask the vendor to meet. In theory, the best lead time is the shortest you can get without paying a premium price to get it. In practice, even paying a premium may not be bad, as long as it can be justified. As they say in a court room, "Don't ask a witness a question unless you already know the answer." The same principle applies to dealing with your suppliers: "Don't ask for a shorter lead time unless getting it will cut the stocking level."

6.2.9 Commingling Overhaul Spares With Safety Stock

Many storerooms operate under the practice of commingling overhaul or outage spares with safely stock for the same item. As a result, demand history tends to get contaminated when calculating safety

Calculating the impact of a 25% lead time reduction before contacting the vendor

Part No.	Description	Part cost ($)	Lead time (Weeks)	Current MIN	Suggested MIN	Savings ($)
830031	MOTOR AC 250HP	19,225.00	8	1	0	19,225
845345	COIL MAIN	3,000.00	12	1	0	3,000
846790	TRIP CARD	2,681.97	10	1	0	2,682
037899	AMPLISTAT X4T	1,160.10	7	0	-1	1,610
569873	RECTIFIER	1,332.73	8	2	1	1,333
337678	BRUSHHOLDER	1,261.00	8	2	1	1,261
435645	BOARD CIRCUIT	1,075.23	15	2	1	1,075
344545	IMPELLER	913.00	8	1	0	913
768577	RELAY VOLTAGE	752.58	7	2	1	753

Total potential savings   31,852

Figure 6.7 Prioritize Lead Time Reductions

stock levels. These two types of inventory should not be mixed, and counter withdrawal records should distinguish whether the issue is for an overhaul project or for an equipment failure. In theory, an overhaul is a planned event with a specific time need for the parts. A failure in service is neither planned nor known in advance. Mixing the two can lead to unnecessary purchases.

6.2.10. Not Considering The Incremental MAX

In Chapter 3, we showed how the Incremental MAX is a nice way to avoid overstocking of expensive items. If the price of an item is high, and the availability improvement of adding another spare is marginal, the incremental MAX value can often exceed $1,000,000 per one percent improvement in availability. Applying the concept of the incremental MAX will decrease the reorder point level and avoid stocking more inventory than necessary.

6.2.11 Failing to Weed Out Duplication

Carrying the same spare under more than one stock number is common in most production storerooms. This problem develops because over the years most plants have no consistent way to describe, or catalog, parts. As new parts are ordered, the assignment of a part identifier is somewhat arbitrary. In Chapter 9, we will cite a case study that addresses this problem.

6.2.12 Buying Too Many Spares Because of Cycle Count Errors

Cycle counting is a necessary practice in storerooms to reconcile the level of inventory shown in computer records with the actual quantity on the storeroom shelf. There are several reasons why the records and actual balances can be different: 1) sloppy practices in documenting distributions, 2) data entry errors, 3) restocking returns to the wrong bin, 4) inaccurate physical counts, and 5) plain old pilfering. Regardless of the reason, inaccuracies between the computer and actual records can often lead to unnecessary purchases, as shown in the following case study.

CASE STUDY 6-6: AJAX STEEL

The Situation:

The storeroom at AJAX Steel just completed its annual one-month cycle count to reconcile actual inventory levels with those shown in the material management system. Under the program, two stock clerks counted the number of items in bins for six hours per day for a thirty day period (1,600 SKUs were counted, approximately 10 percent of the total items). The results showed a 4.7 percent discrepancy error. The accounting and storeroom manager were discussing the results with the purchasing manager, who commented, "I wonder how many of those higher count items have hit the reorder point in the computer and were purchased prematurely?"

The Proposed Solution:

The storeroom manager agreed to collect the data necessary to answer the question.

The Numbers:

Figure 6.8 shows the results of the review. A total of 75 records had actual balances different than the computer records: 18 higher and 57 lower. The purchasing records indicated that five of the 18, all slow-movers, had been replenished during the last six months. After looking at the usage history for each, the storeroom manager estimated the present worth of the unnecessary purchases to be $607 at the company's 12-percent carrying cost rate.

The Conclusion:

Although not a trivial amount, the $607 premature purchasing cost was not considered sufficient reason to cycle count more frequently.

Five of 18 discrepancy items showing actual balance higher than computer balance

Part No.	Description	Unit price ($)	Cycle count amount	Computer amount	Difference	Value of difference
S87995	Pressure plug	1,602	4	3	1	$1,602
S09223	Circuit board	2,016	2	1	1	2,016
S11232	Clamp	474	6	4	2	948
S55355	Pressure value	961	2	1	1	961
S14444	Gasket	87	12	11	1	8 7

Total $5,060

Figure 6.8 The Impact of Cycle Count Errors

6.3 SOME ADDITIONAL THOUGHTS ON PURCHASING

6.3.1 Resources

Thus far, we have talked a lot about the consequences of buying spare parts before they are needed. At the other extreme is the problem of purchasing too late and having to expedite. The purchasing function is not a pure science; despite all the advances in computers and software codes to optimize stocking, it remains still pretty much a people-driven function. Certainly purchasing departments have not been immune lately to cost cuts and personnel reductions seen throughout industry. With fewer resources available to handle the same or increasing work loads, things inevitably get delayed. The old saying that "time is money" is certainly true when material and spares are being purchased.

6.3.2 Purchasing Procedures

People have a bad habit of not following verbal instructions. Therefore, if you expect consistent adherence to certain tasks, it better be in writing. This is especially true for purchasing procedures. Our work with clients has shown that the following purchasing actions should be documented in the purchasing procedures:

- When to call a supplier for a price update.
- When to ask a supplier for a current lead time.
- When to require justification from the requisitioning person.
- Who can approval a vendor shipment delay.
- What support information is required to accompany a requisition.

To the extent possible, the above issues should be imbedded in the purchasing procedures (electronic or hardcopy) to insure consistent compliance.

Circuit Breaker

CHAPTER 7

UNIQUE SOLUTIONS TO EVERYDAY INVENTORY PROBLEMS

7.1 WHAT THE READER SHOULD LEARN FROM THIS CHAPTER:

- Where to focus your inventory management program
- How to solve a queuing situation
- Unique solutions to everyday inventory problems

7.2 WHERE TO FOCUS YOUR INVENTORY MANAGEMENT EFFORTS

Figure 7.1 lists eight areas where significant benefits can result from focusing your inventory management efforts. Listed beside each area is a reference to a chapter in this book where the subject is described in more detail. In the remainder of this chapter, we will cover numerous case studies that impact on one or more of these areas.

7.3 SELECTED CASE STUDIES

Solutions for many everyday inventory management problems will be discussed in this section. Table 7.1 identifies problems and their potential solutions.

Table 7.1

Problem Needing A Solution	Proposed Solution
Overload at storeroom counter	Queuing analysis
Reduce unemployed assets	Avoid replenishing excess inventory
Short-term availability problem and cash shortage	Adjust no-cost understocks only
Need to trim long-term assets	Conduct After 4-Year analysis
Recent unexplained inventory increases	Review top 10 reports
How to set realistic goals for next year	AVR what-if analysis
Identify inventory imbalance	Perform cost bias analysis

1. Improve the asset base by reducing overstocked items (Chapter 5).

2. Lower the risk of running out by increasing understocks (Chapter 3 and 7).

3. Identify and dispose of obsolete items (Chapter 5).

4. Dispose of excess items that will not work off (Chapter 5).

5. Partner with vendors/distributors (Chapter 9).

6. Stock common spares centrally (Chapter 5).

7. Improve lead times with suppliers (Chapter 5).

8. Upgrade your procurement procedures (Chapter 6).

Figure 7.1 Focus Your Inventory Management Program

7.3.1 Queuing Situations

Queues occur in many situations such as fast food restaurants, bank teller windows, theater ticket cashiers, and parts counters at storerooms. In most cases, arrivals for service in these situations have been shown to follow a Poisson distribution, as discussed in Chapter 2. For that reason, many of the formulas and probability values are routinely tabulated in standard tables for easy use when performing calculations for queuing situations. Factors that come into play in evaluating a queue are:

A = Average arrival rate of users to the counter

S = Average service rate (users served per hour)

M = Number of service lines in operation

Cu = Typical wage/hour of user

Cs = Wage/hour of counter service personnel

N = Average number of users in system (waiting and being

　　served) (Values derived from standard tables)

T = Idle time per hour of operation (values derived from tables)

W = Average number of customers waiting in line

T = Average waiting time(Hours)

This definition of terms and the corresponding values for certain terms were obtained from Colley et al, *Operations Planning and Control,* Holden-Day, (1978).

CASE STUDY 7-1. AJAX MANUFACTURING

The Situation:

AJAX Manufacturing is a midwestern U.S. company engaged in the production of medium and heavy-duty air compressors for industrial duty. Recently the company made a $7 million investment to upgrade and install a new assembly line for manufacturing its new line of high-performance air compressors. Unfortunately, since the line was started up, there have been an unusually high incidence of failures, especially bearing and pinion gears. Adding to the difficulty has been heavy congestion at the storeroom service counter when maintenance personnel arrive to retrieve replacement parts.

The storeroom manager explained the problem this way, "In the past, we have always been able to service maintenance with only one counter clerk, but over the last month or so we have often had up to seven or eight people waiting in line to get the parts they wanted. Complicating the situation is the fact that the equipment bills of material have not been digitized and computerized. This slows things down when our clerk and maintenance try to figure out exactly which spare is needed to replace the failed one." The maintenance superintendent comments on the problem were more blunt, "I don't care what it takes to fix the problem! All I know is I can't afford to have half my maintenance workers waiting in line for up to half hour at a time."

The Proposed Solution:

Two solutions were proposed: 1) adding a second storeroom attendant on peak days, and 2) improving the long-term efficiency of the storeroom by adding automatic picking equipment.

The Numbers:

Random arrivals of users at the counter varied from 2 to 11 per hour, with a weighted average of 8 per hour. At peak periods, usually from 7:00 to 9:00 a.m. and from 1:00 to 2:00 p.m., as well as unpredictable occasions, *queues were forming, causing delays to the users who earn an average of $20.00/hr (storeroom attendants average $12.00/hr). Other supporting data were:*

$$M = 1 \text{ (single-channel service)}$$
$$A = \text{Average arrival rate} = 8 \text{ per hour}$$
$$S = \text{Average service rate} = 5 \text{ min} = 12 \text{ per hour}$$
$$A/SM = (8)/(12)(1) = 0.67$$

Typical table for *W* (average number of customers waiting in line)

A/MS	M=1	M=2	M=3	M=4	M=5
0.60	0.9000	0. 6743	0. 5313	0.4295	0.3527
0.65	1.2071	0.9653	0.7782	0.7362	0.5550
0.70	1.6333	1.3298	1.1301	1.0258	0.8523

Typical table for *T* (average waiting time)

0.60	0.4000	0.8000	1.2000	1.6000	2.0000
0.65	0.3500	0.7000	1.0500	1.4000	1.7500
0.70	0.3000	0.6000	0.9000	1.2000	1.5000

By interpolation, W = 1.37
                    T = 0.33
(C*u*)(W) = ($20.00)(1.37) = $27.40
(C*s*) (T) = ($12.00)(0.33) = $3.96

> Total idle cost is
> $31.36/hr of which
> 87% is by users

Figure 7.2 Calculating W and T

Figure 7.2 summarizes the determination of W (average number of customers waiting) and T (average waiting time) from standard lookup tables, using interpolation. Results indicate that W = 1.37 and T = 0.33.

Multiplying these values by the user and attendant wage rates yields a total idle cost of $31.36/hour, of which 87 percent is for the users. Figure 7.3 summarizes the total idle cost over the range of user arrival rates. A look at Figure 7.4 shows that the total idle time increases significantly as the average arrival rate of the users approaches 11 per hour.

Cu = $20.00 Cs = $12.00

A	S	M	A/MS	W	T	CuW	CsT	$CuW+CsT$
2	12	1	0.17	0.03	0.83	$0.60	$9.96	$10.56
4	12	1	0.33	0.15	0.67	$3.00	$8.04	$11.04
6	12	1	0.50	0.50	0.50	$10.00	$6.00	$16.00
8	12	1	0.67	1.37	0.33	$27.40	$3.96	$31.36
10	12	1	0.83	4.17	0.17	$83.40	$2.04	$85.44
11	12	1	0.92	12.10	0.08	242.00	$0.96	$242.96

Base Case

Figure 7.3 Total Idle Cost Versus User Arrival Rates

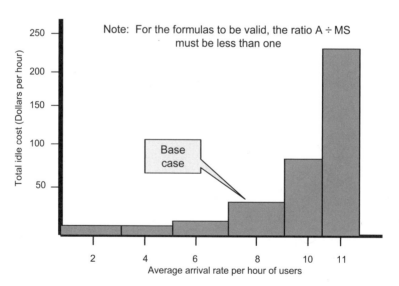

Figure 7.4 Changes in Total Idle Time

Options to improve storeroom situation:

Add a Second Storeroom Attendant on Peak Days:

Total Idle Time ($C_uW + C_sT$) = $ 17.92, a savings of $13.44 per hour, which is more than the hourly wage of the extra attendant

Improve the Efficiency of the Storeroom:

Estimates indicate that automatic picking equipment and bin reorganizing, with a cost of $150,000, would lower the *average service rate* from 5 to 3 minutes. This savings of $18.72 per hour would give a payback on the investment of about 4 years.

Figure 7.5 Summarizing The Two Options

The Conclusion:

Figure 7.5 summarizes the two options considered to solve the problem. Adding a second storeroom attendant on peak days was shown to save $13.44 per hour, $1.44 per hour more than the average cost of adding the extra attendant.

For the extra investment of $150,000 to add some automatic picking equipment and reorganizing the storeroom, a savings of $18.72 per hour was achieved. The payback on the investment was determined to be about 4 years.

The plant management elected to add the extra attendant.

7.3.2 Writing Off Excess Inventory

In Chapter 1 we discussed how inventory fits into the broader issue of return-on-investment. We noted that there are only two ways to dispose of excess inventory: work it off or dispose of it. To dispose may require the company to take a hit on the bottom line unless they happen to have some tax loss carryovers available. In this next case, the company does.

Valve Guides

CASE STUDY 7-2. AJAX STEEL

The Situation:

AJAX steel is a medium-sized U.S. steel company operating two rolling mills, one in Alabama and one in Indiana. Although the overall outlook for domestic steel improved over the last few years, AJAX was still overburdened with excess capacity and higher-than-average costs to manufacture. The new CEO recently instituted a major cost reduction program that targeted improving the return-on-investment (ROI) by 10 percent per year over the next five years. All aspects of the ROI equation were addressed, including cost of goods sold (labor, material, and factory overhead), administrative expense, and permanent and current assets, including inventory. The company's chief financial officer explained the program this way, "We have way too many assets tied up in inventory at our two mills. We're still carrying inventory at levels that would support double our anticipated long-term capacity projections. We need a sensible program to get it down. For long-range planning purposes, I need to have some firm numbers on what's do-able".

The Proposed Solution:

An analysis of the total inventory at each mill was proposed. Inventory was divided into three categories: raw materials and supplies, work-in-progress, and finished goods. Teams were assigned to work on a specific category. The main focus of the work-in-progress investigation was a thorough review of the spare parts situation, which was considered excessive.

The Numbers:

The analysis of the inventory accounts at each plant revealed the following:

Inventory Category	Georgia Plant	Indiana Plant
Active Inventory	$2,106,000	$1,872,000
Rarely-Used Inventory	$19,620,000	$18,485,000
Commodity Inventory	$1,654,000	$1,355,000
Total	$23,380,000	$21,712,000

Turnover on active and commodity inventory was determined to be 7.1 and 6.4 turns, respectively. Turnover for the rarely-used inventory was only 0.31 turns. Because turnover was not considered to be a valid measure for monitoring rarely-used inventory, it was ignored. Instead, a decision support tool for slow-moving inventory was used to estimate the expected work-off of spares each year for the next four years. These results are shown in Figure 7.6. First, $17,868,000 (46 per cent) of the total $38,105,000 rarely-used inventory was determined to be overstocked to meet mill availability objectives. Of that amount, about one-half ($8,964,000) was expected to work off through usage over the next four years. Because the company had over $10,000,000 of tax-loss carry-forwards available to apply against inventory write-downs, the CFO prepared a proposal to the CEO to apply some of the tax-loss credits to writing off excess inventory now.

The Conclusion:

After reviewing all potential uses of the tax-loss credits, the management wrote off all of the excess inventory remaining after four years ($8,964,000).

(All values in $000ís)

Store	Total inventory	Overstocks	Cash flow savings based on anticipated usage				
			Year 1	Year 2	Year 3	Year 4	Total
Georgia	19,620	10,072	2,532	1,369	1,033	585	5,519
Indiana	18,485	7,796	1,401	970	680	394	3,445
Total	38,105	17,868	3,933	2,339	1,713	979	8,964

|←—— 50% of overstocks will work off ——→|
over first 4 years

Figure 7.6 Estimating Expected Work Off of Spares

7.3.3 Fixing Understocked Items

Back in Chapter 2, we talked about the cost bias, the tendency to overstock cheaper spares and understock expensive ones. Stocking short is the dread of maintenance personnel because they see it as hampering their ability to keep the equipment running. But fixing re-order points that are set too low to support desired availability does not always require buying additional inventory, as the next case study will show.

Manway Gaskets

CASE STUDY 7-3. AJAX REFINING

The Situation:

AJAX Refining is a 320,000-barrel-per-day refinery in the Houston area producing a variety of petroleum products. The plant was originally erected in 1967 and has had a few major upgrades to the process over the years. Because of its age, the recent owners were able to purchase the plant at a bargain price. They expect to yield a 16 percent return-on-investment if plant operating expenses and capital expenditures can be held to current levels. But because of the age of much of the equipment, breakdowns have shown an upward trend; backorders of replacement spare parts have caused a 6 percent loss of capacity over the last year.

The refinery maintenance superintendent put it this way, "I think we're doing a pretty good job of keeping this place running, but it seems like every time one of my workers goes to the storeroom to get a replacement spare they don't have any and it's on backorder. For a lot of the stuff it doesn't matter, but if the part is critical to maintaining production we're hurting. I don't mind buying more spares if we really need them and the only way I know how to fix the problem is just tell the storeroom to increase all reorder points 1 or 2 units, but I know that will cost us a bundle. Besides, my budget to buy inventory is almost depleted anyway. There must be a better way to reverse the backorder trend."

The Proposed Solution:

At the next plant manager's meeting, the supply-chain manager suggested hiring some outside consultants familiar with production inventory. This suggested was approved. After reviewing a data file of the refinery inventory, the consultants discussed their recommendations.

Because of the limited amount of maintenance resources available to participate in the program, a low-resource, conservative approach was adopted to solve the problem. First, the inventory was sorted into its components using the inventory tree approach (see Chapter 1). Next, safe default parameters were set both for part criticality and for lead time to replenish; these, along with recent usage history, allowed a MIN/MAX to be determined for each rarely-used item. Suggested new MIN/MAXs were then compared to current material management values and the parts divided into three categories: 1) likely overstocks, 2) likely understocks, and 3) those that required no MIN/MAX adjustment because the current stock levels matched the conservative suggested values.

The Numbers:

Of the 26,450 rarely-used stock items in the storeroom, 13,440 were considered overstocks and were not adjusted because the concern was to focus on understocks. A total of 3,780 understocks were identified that required the reorder point to be raised one or more units to meet the high criticality availability minimum (99.9 per cent) used in the analysis for each item. Of these, 2,210 could have their reorder point raised without incurring any cost because the current balance-on-hand was high enough to avoid an immediate replenishment order. Further analysis showed that a whopping 81 per cent of the backorders during the last year could have been avoided if the recommended increases in reorder points had been adopted earlier.

Figure 7.7 summarizes this approach for a sample of four parts. To meet the minimum 99.9 per cent availability requirement, part 70216 needed to have the current minimum raised from 1 to 2 units, thereby increasing its availability from 99.85 to 99.99 per cent. Because there were already 12 on the shelf, ten more units would have to be used to trigger a replenishment order. Similarly, part 70344 needed to be raised from 0 to 1; part 76651 from 1 to 2; and part 73844 from 3 to 7 units. But only 5 units of part 73844 were on the shelf; therefore, its MIN could only be raised from 3 to 4 units without triggering a reorder. That increase only raised the availability to 97.32 per cent; to go higher would require a purchase.

The Conclusion:

After reviewing the consultants' recommendations, the decision was made to raise the reorder points for the 2,210 no-cost understocks to the recommended level. In addition, the remaining 1,570 cost understocks were reviewed and 93 had their reorder point raised, requiring a total purchase requisition of $117,453.

The Balance-on-Hand determines if an understocked spare needs to be replenished immediately

Part	Price (Dollars)	Balance-on-hand	Suggested MIN	Current MIN	Current Avail (%)	Increase in MIN From – To		New Avail (%)
70216	396.00	12	2	1	99.85	1	2	99.99
70344	1,520.10	2	1	0	98.62	0	1	99.99
73844	2,749.20	5	7	3	92.54	3	4	97.32
76651	51.68	43	2	1	98.96	1	2	99.95

The MIN for part 73844 cannot be raised higher than 4 units without buying additional inventory.

The objective of the program was to raise the availability of all items to at least 99.9%. Without buying more inventory part 73844 could not meet the objective.

Figure 7.7 Improving Availability Of Understocks

7.3.4   When to Use Investment Recovery

We talked about investment recovery in Chapter 5, showing that powerful economic tools are available for deciding how much excess inventory should be disposed. This next case study for AJAX Paper illustrates how an everyday problem can be solved using investment recovery techniques.

CASE STUDY 7-4 AJAX PAPER

The Situation:

AJAX Paper is a well-established Canadian paper company that markets a variety of paper products including fine paper, linerboard , and an assortment of packaging products. Competition is fierce in all of its markets, especially for packaging products. Although the economy for the paper industry has been good for the last three years, the industry usually goes through an up-and-down business cycle about every 4-5 years.

The company's chief financial officer explained the situation this way, "We've had to make some pretty heavy investments in plant and equipment lately to keep up with the recent market boom. Our asset interest charges are beginning to push allowable limits. I anticipate a mild recession in the next year or so, and we've got to start shedding assets over the next few years to get thing back to where we want them. I know we have a lot of excess inventory and I argue about it all the time with the production manager. Their philosophy is 'since we've already bought it, why write it off?' I know there are some financial advantages to getting rid of some of it and I want to present a case to the VP. I want to write off all the excess we can't work off over the next 3-4 years. I need numbers."

The Proposed Solution:

The CFO needed two things: 1) a good determination of which specific stock keeping units (SKUs) were expected to be excess four years from now, and 2) a prioritized listing of each item, showing how many units were excess. Both of these would be needed to convince the operations manager to agree to the write off. The CFO gathered the data using a decision support tool specifically designed to handle slow-moving spares. Each SKU with an inventory value of over $500 was analyzed and the excess determined. The amount remaining after four years was then estimated from past usage rates and the prioritized lists prepared.

The Numbers:

A list of 347 stock items was determined, totaling $6,255,000 of anticipated excess after four years. Each item was carefully reviewed by a team consisting of maintenance, materials, and controller personnel, who were able to agree that all but 61 of the items could be eliminated. Figure 7.8 shows five of the items on the list. Decisions about any of the 61 remaining items were left to the operations manager and the CFO.

The Conclusion:

After considerable debate lasting over several meetings, the operations manager agreed to write off 37 items totaling $416,500.

Part No.	Current value of overstocks (Dollars)	Anticipated usage next 4 years (Dollars)	Overstocks after 4 years (Dollars) (Priority order)	Units after 4 years
623F84	25,000	2,380	22,620	9
78552A	11,288	512	10,776	20
675554	222,185	211,594	10,591	0
143452	9,900	393	9,507	23
908788	4,965	39	4,926	298

Figure 7.8 Analyzing Excess Inventory

7.3.5   Keep An Eye on Price and Quantity Fluctuations

The following case study will highlight a simple, but effective way to keep an eye on price or quantity fluctuations through Top 10 reports. These reports employ the Pareto rule (the top 10 items make up much of the volatility swing) and the principle of prioritizing the items from highest to lowest impact.

CASE STUDY 7-5 AJAX CHEMICAL

The Situation:

AJAX Chemical is a medium-size New Jersey supplier of chemicals to the paint, refining, and plastics industries. A recent downturn in many of their markets forced the company to cut personnel and close several process lines. Budgets have been cut and several managers have been fired for not operating within budget. Everybody got the message–make budget or your gone! The plant manager at the Long Branch plant put it this way, "Things are so tight around here, I spend all of my time huddled with the controller looking at every way possible to cut costs. My main concern is that we may be 'going on the cheap' with our maintenance program, and it will come back to haunt us over the next year or so. What I need is some good reports that will allow me to keep a close eye on costs to keep them in line with budget."

The Proposed Solution:

One area that has been showing volatility from week-to-week has been the inventory stock account. Major swings in inventory value were occurring. The supply-chain manager, who was personally reviewing every requisition over $100 before it was released, decided that a timely, Pareto Rule-type report was needed, one that would focus on the few items that were causing most of the cost volatility. The manager's assistant came up with a set of Top 10 reports, shown in abbreviated format in Figures 7.9 through 7.12, for both increases and decreases in SKU price and quantity.

The Numbers:

The most recent lists for the last week showed the Top 10 inventory increases due to price changes (Figure 7.9) ranging from $14,801 to $1,868. With all the effort lately to control procurement, a couple of the numbers looked out-of-line, especially part number 20472. On further review, it was determined that there had been a data entry error in the price for Week 15; the latest price should have been $5,403, not $11,503. In addition to allowing the inventory value to be adjusted, the report was useful in catching date entry errors that would have gone undetected until the next receipt for the item.

Increase = Week 15 quantity times (Week 15 cost ñ Week 14 cost)

Stock rank	Part No.	Week 14 cost (Dollars)	Week 14 on hand	Week 15 cost (Dollars)	Week 15 on hand	Inventory increase ($)
1	20472	4,100.00	1	11,500.34 *	2	14,801
2	20496	588.00	2	2,029.50	4	5,766
. . . .	. . . .	. . . .	. . . .	. . . .	. . . .	. . . .
9	23245	680.00	1	2,064.00	2	2,768
10	54444	206.40	6	439.95	8	1,868

* Week 15 price for part 20472 should have been $5,403.00.

Figure 7.9 Top 10 Increases In Spare Part Price

Increase = Week 14 cost times (Week 15 quantity ñ Week 14 quantity)

Stock rank	Part No.	Week 14 cost (Dollars)	Week 14 on hand	Week 15 cost (Dollars)	Week 15 on hand	Inventory increase ($)
1	21472	13,440.00	2	13,760.00	4	26,880
2	28996	16,483.00	0	16,483.00	1	16,483
. . . .	. . . .	. . . .	. . . .	. . . .	. . . .	. . . .
9	41245 *	4,060.00	0	2,585.00	2	8,120
10	54119	895.29	3	918.97	10	6,267

* Notice the price for part 41245 went down in week 15.

Figure 7.10 Top 10 Increases In Spare Part Quantity

The Conclusion:

By looking at the Top 10 reports, the plant management was able: 1) to identify significant swings in price and quantity for items, 2) to correct data entry errors quickly, 3) to adjust order quantities when the amount purchased seemed excessive, and 4) to conduct cycle counts on selected items if the balances-on-hand seemed unreliable.

Decrease = Week 15 quantity times (Week 15 cost ñ Week 14 cost)

Stock rank	Part No.	Week 14 cost (Dollars)	Week 14 on hand	Week 15 cost (Dollars)	Week 15 on hand	Inventory decrease ($)
1	26672	2,920.00	0	1,436.67	3	4,450
2	27896	6,206.20	1	4,351.23	2	3,710
....	....	....	....	....	....	....
9	23445	2,862.99	1	2,384.00	1	479
10	59849	1,000.00	1	763.64	1	236

Figure 7.11 Top 10 Decreases In Spare Part Price

Decrease = Week 14 cost times (Week 15 quantity – Week 14 quantity)

Stock rank	Part No.	Week 14 cost (Dollars)	Week 14 on hand	Week 15 cost (Dollars)	Week 15 on hand	Inventory decrease ($)
1	26882	6,225.00	2	6,225.00	0	12,450
2	27406	1,373.45	17	1,373.45	8	12,361
....	....	....	....	....	....	....
9	28945	5,067.70	1	5,067.70	0	5,068
10	59559	5,013.21	1	5,013.21	0	5,013

Figure 7.12 Top 10 Decreases In Spare Part Quantity

7.3.6   Monitoring Performance Against Goal

We will discuss goal setting and monitoring performance in more detail in Chapter 8. However, the following case study for AJAX Utility demonstrates a unique tool, the AVR, for tracking performance against goal. It is especially unique because the AVR can be used to quantify future planned actions against goal and predict their effect on the goal.

CASE STUDY 7-6 AJAX UTILITY

The Situation:

AJAX Utility was a mid-size independent power producer who over the last three years added generating plants in Nevada and Texas to its existing plants in Kansas and Michigan. All together, the utility now had about 1,450 megawatts of generation burning a mixture of fuels including coal, gas, oil, and biomass. All of the plants were peaking units, meaning they generally operated most often during peak load periods or when larger base-load plants were off-line for maintenance or other reasons.

The company VP was a stickler for setting goals for anything that could be measured, operating on the philosophy that "people respond to what you inspect, not what you expect. Most of the inventory assets at the stations were tied up in rarely-used inventory, which did not move very much. As a result, turnover couldn't be used as a metric to monitor spare parts inventory performance. What the VP wanted was, "some measure of performance that will tell me if we are carrying the right amount of inventory at our stations; not too much, nor too little. It would also be nice if I could evaluate in advance if certain actions on our part will really improve our inventory situation, and by how much."

The Proposed Solution:

At a recent conference, one of the station managers heard a presentation by a consultant who professed to have a methodology for setting and monitoring rarely-used inventory, something called the "Absolute Variance Ratio (AVR)." The decision was made to invite the consultant to headquarters to discuss the AVR concept. Subsequently, the utility contracted with the consultant, supplying a data file of the inventory at the stations. With support from various station operating and management personnel, the consultant recommended goals for the forthcoming year. (For a more thorough discussion of AVR see Chapter 8.)

The Numbers:

After analyzing the data from each plant, the consultants prepared Figure 7.13. It shows starting AVRs ranging from a high of 0.81 to a best (lower is better) of 0.46. Normal starting AVRs typically run in the range of 0.50 to 0.80; anything over 1.00 shows significant imbalance in the stocking of spares. Figure 7.13 also shows the results of a what-if analysis: 1) what would happen to the AVR if the plants would agree to lower 4 out of every 5 stocking MINs from the current value to the recommended value, 2) what additional improvement in AVR would be possible if all the no-cost increases were made for understocks, 3) what further improvement in AVR would be possible if all current order quantities were changed to the economic order quantity(EOQ) amount, and 4) what the final AVR would be if all understocks were increased to recommended values by making the necessary purchases. The column shown in gray shading is the best (lowest) AVR possible under the scenario without actually buying more inventory. These values became the target AVRs. If the AVR targets were achieved, the Nevada plant would be one of the only plants worldwide to ever achieve an AVR less than 0.15.

The Conclusion:

The AVR targets for the next eighteen months were approved and became a portion of the ratings of the various station management for bonus purposes. Although all stations made excellent improvement, only the Kansas and Nevada stations beat the goal, as shown in Figure 7.14.

Plant location	Starting AVR	What the AVR would be if these actions were taken . . .			
		Lower 4 of 5 MINs currently too high	Made all of the ino costî fixes	Adjust all OQs to EOQ	Make all iwith costî fixes
KN	0.74	0.52	0.43	0.40	0.09
MI	0.67	0.43	0.33	0.26	0.13
NV	0.46	0.22	0.17	0.11	0.04
TX	0.81	0.66	0.49	0.39	0.16

This column shows the lowest AVR achievable without buying more inventory

The target AVR

Figure 7.13 Setting Goals With The Absolute Variance Ratio (AVR)

7.3.7 Fixing the Cost Bias

We explained earlier that everybody has a cost bias in how they stock their spare parts inventory. But before you can fix the problem, you have to know how bad it is. The next case study focuses on this issue.

Plant location	Starting AVR	Change in AVR by quarter					
		1	2	3	4	5	6
KN	0.74	0.51	0.50	0.46	0.46	0.41	0.39
MI	0.67	0.51	0.47	0.44	0.41	0.39	0.36
NV	0.46	0.42	0.31	0.27	0.22	0.11	0.08
TX	0.81	0.62	0.55	0.57	0.54	0.50	0.47

Only the KN and NV stations met their AVR goal of 0.40 and 0.11, respectively.

Figure 7.14 Measuring Results With AVR

Part cost ($)	Number of Items	Recommended inventory ($)	Current inventory ($)	Ratio of current to recommended	% Items understocked
< 100	112	24,588	86,036	3.5	17.9
100 < 250	254	136,956	263,788	1.9	34.6
250 < 500	303	218,622	388,007	1.8	29.7
500 < 1,000	326	402,189	588,200	1.5	30.7
1,000 < 2,500	260	679,991	872,921	1.3	31.2
2,500 < 5,000	71	389,206	407,237	1.0	32.4
5,000 < 10,000	30	367,797	388,377	1.1	40.0
> 10,000	20	748,600	608,303	0.8	45.0
Totals	1,376	2,967,949	3,602,869	1.2	30.7

The Cost-Bias

Figure 7.15 Analyzing The Cost Bias

CASE STUDY 7-7 AJAX METALS

The Situation:

AJAX Metals is a small company in the midwest producing stainless steel tubing and fittings for the general industry. Their only plant has a relatively modern foundry and a state-of-the-art continuous casting line for making tubing in a variety of sizes and wall thicknesses. The new supply-chain manager, who was recently hired from the aluminum industry after spending 12 years as material manager in a plant making engine blocks, had heard about the so-called cost bias in the stocking of slow-moving inventory, and was curious to see how bad or good the situation was at the AJAX plant. The following were the supply-chain manager's thoughts, "I know I have a cost bias, because I understand every plant has one. Before I can do much to fix it, I need to know how bad it is– severe, modest, or minor. I've been allowed to bring in some consultants who will do an initial analysis, at no charge, to see how bad my situation is. All I need to do is provide them some data. Not a bad deal!"

The Proposed Solution:

The solution was a "no brainer." All the plant management had to do was provide a data file containing usage data for the last three years, part prices, lead times, balances, and current stocking levels. The consultants did all the rest.

The Numbers:

Figure 7.15 shows the results of the cost bias study. The chart below compares those results to norms for severe, normal, and mild biases:

Bias Level	Range of Cost Bias
Severe	Over 12 for <$100 parts to
	Under 0.5 for >$10,000 parts
Normal	8 to 0.7
Mild	4 to 0.9

Fortunately for AJAX Metals their cost bias was in the mild range, 3.5 for low cost parts to 0.8 at the high end. Even though the bias was mild, there were still fairly high percentages of items understocked in each part cost range. Overall, about 30 percent of the 1,376 SKUs were understocked.

The Conclusion:
Money was budgeted to hire the consultants to provide detailed recommendations for adjusting MIN/MAXs to fix both overstocks and no-cost understocks. The adjustments were made over the next three months, allowing the AVR to decrease from 0.62 to 0.30.

7.3.8 Losing An Opportunity

In Chapter 6, we discussed the opportunity to avoid unnecessary purchases by knowing when to defer reorders for spares that have too high a current reorder point. Stopping front-end purchases is one of the few opportunities available to plant management to save cash now (this applies to both replenishments and the purchase of new spares). Failure to take advantage of knowledge that can stop unnecessary purchases is clearly a lost opportunity to save money. The next case study shows that at least one manager is interested enough to quantify the opportunity.

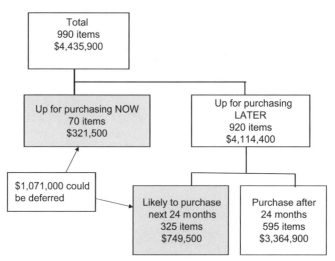

Figure 7.16 The Cost Of Premature Purchases

CASE STUDY 7-8 AJAX FERTILIZER

The Situation:

AJAX Fertilizer operates a processing plant in Texas City, Texas, to manufacture fertilizer for the agriculture market. Recently they hired a consultant to help set stocking levels for their safety spares. After the project was over and the recommended changes to current MIN/MAXs had been presented, there was "push-back" from some of the maintenance superintendents to implement the recommended changes. The supply-chain manager asked the consultants to help justify the impact of not making the changes, "Can you quantify the lost opportunity, in dollars, from buying purchases prematurely over the next year or two if we don't make the MIN/MAX changes? I want to show how much we can defer by fixing these stocking levels."

The Proposed Solution:

Having previously analyzed the plant inventory data, the consultants only had to consider two issues: 1) which stock items were determined to be overstocked, and 2) which ones were likely to come up for replenishment in the next twenty-four months. They set out to complete the analysis.

The Numbers

Only 990 items were currently under discussion for review (see Figure 7.16). Of these, 70 items were currently in the purchasing process, and another 325 were projected to come up for repurchase over the next two years if usage continued at the same rate as during the last two years. According to the consultants, all of these 395 items had a current reorder point that was higher than necessary to meet the availability requested by maintenance. Lowering the MINs for the items could avoid spending over $1 million prematurely. Another $3.4 million opportunity awaits after 24 months.

The Conclusion:

After seeing the cost avoidance opportunity, the plant manager issued the following instruction to maintenance, "Make the suggested changes to the reorder points or give me a written reason why you think you shouldn't. Cash is short, and we don't need to spend it before we have to."

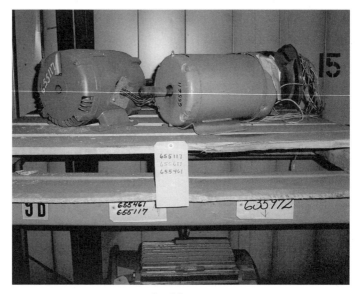

5hp Motors

CHAPTER 8

SETTING AND MONITORING GOALS

8.1 WHAT THE READER SHOULD LEARN FROM THIS CHAPTER:

- Why bother to set goals
- Suggestions for setting goals
- Deciding what to measure
- Why turnover is not a good measure for slow-moving spares
- Why AVR is a better measure for slow-movers
- Unique solutions to everyday problems

8.2 WHY SET GOALS?

Most people have goals in life. Many goals are achievable, others are not: owning your own company someday (probably achievable), becoming a millionaire (reasonably achievable), being elected President of the United States (unlikely). With personal goals, there's generally no great penalty if you fail to achieve them, maybe just some disappointment. The same is not true in business. Your salary, your bonus, and very likely your job depend on meeting your piece of the company goals.

Why do companies set goals? The simple answer is they're expected to by the owners (shareholders). Goals also motivate people, especially when there are rewards for achieving them. Without goals, many people would believe they deserve lifetime employment just for showing up each workday.

8.3   SETTING GOALS

In this section we will focus our attention on setting goals for the company. Understand that not all goals are equal in importance. Therefore, companies frequently sort goals into primary and secondary goals. Primary goals should concentrate on those key metrics that impact the financial performance of the company. These include sales, market share, return-on-investment, and share price.

Secondary goals can be selected to concentrate on areas where the company needs to improve performance if the primary goals are to be achieved. The secondary goals set for one organization are often not appropriate for another organization. For example, if the sales organization had recently experienced high personnel attrition, it might need a vigorous new recruitment and training program as a key secondary goal. If shop productivity traditionally slips prior to labor contract negotiations, a secondary goal may be needed by the manufacturing organization to offset or avoid productivity deterioration. Affirmative action goals to hire minorities might be set by all units.

8.3.1-Dual Level Goals

Experience has shown that it is useful to set two levels for each goal, a commitment level and a target level. The commitment level should be set at that level of performance necessary for the department or profit center to meet its committed level of performance to higher management. It has been suggested that the probability of achieving the commitment-level goal should be high, about 95 percent. If most units meet their commitment level of performance, then the company would meet its commitments to the stockholders and the financial analysis, who have a major impact on the stock price.

The target level should be set higher than the commitment level for all goals. Doing so: 1) establishes a stretch for achievement purposes, 2) sets an upper benchmark on the performance evaluation scale, and 3) serves as a basis for setting next years commitment level. It has been suggested that the target-level goal be set so that the probability of achieving it be about 70 percent.

8.3.2 Goals Should Be Quantitative

Qualitative goals just don't work. Setting a goal such as "improve inventory performance" is worthless. How would you know what result is expected? By when? And how would performance be measured? Many goals may seem qualitative, such as "winning the game," even though you haven't stated by how many points. But at least in most sporting events, there is a quantitative basis for knowing who won. An excuse often made for qualitative goals is that "there is no way to measure this." Yet seldom is that true.

There is no simple solution to the problem of setting the quantitative value desired for each primary and secondary goal. Usually, a thorough analysis of part performance is needed, coupled with judgment as to what constitutes a fair and reasonable degree of improvement from current performance. For example, a 20 percent improvement in sales might be a reasonable goal if the market is recovering from a recession. It might not be reasonable if you are predicting a recession.

8.3.3 Monitoring Results

A convenient method for monitoring results to is to establish a Control Chart. Each chart typically tracks the performance of a specific goal, showing the attributes to be measured along with monthly or quarterly performance against goal. For most slow-moving production inventory goals, we have found that quarterly tracking is adequate, whereas for active inventory, monthly or even weekly tracking is appropriate.

8.3.4 Shared Goals

Because setting and monitoring goals usually lead to appraising someone's performance, it must be recognized that no single person has the sole responsibility for achieving most goals. Suppose a goal has been set to lower the internal time for placing and receiving shipments by 20 percent for all stock units. In all likelihood, purchasing, stores, maintenance, administration, and shipping/receiving will all play a

Use a *constant dollar* basis when setting goals and monitoring results. Goals can then be measured without the impact of price inflation or deflation masking the actual improvement.

October 2003 <u>Actual</u> Inventory $ 3,097,206

      Price Impact (+) $ 128,804
      Price Impact (-) $ (39,941)
      Net Impact $ 88,863

August 2004 <u>Adjusted Actual</u> Inventory $ 3,008,343
      (constant dollar basis)

Figure 8.1 Neutralize Price Changes When Monitoring Inventory.

part in setting and achieving that goal. In addition, the fact that goals are shared does not automatically mean that each person must be evaluated or rewarded equally.

8.3.5 Factor Affecting Goal Results

When setting a goal, remember that the results can be impacted by factors that are outside the control of the people being measured against the goal. Suppose a goal has been set to lower the amount of spares inventory for valves by 12 percent by the end of the year. All of the following factors could affect the chances of meeting the goal:

- Prices for the valves could have increased during the year as new purchases are made. For this reason it is best to set the goal on a "constant dollar" basis. Figure 8.1 illustrates this.
- Purchase quantities (the EOQ) may have changed for some items since the start of the year as either price, demand, carrying cost, or cost-to-place the purchase order change. This type of impact is non-controllable and needs to be neutralized when measuring the goal.
- The Average Inventory value for the item may have changed throughout the year. Figure 8.2 explains why it is best to base the goal on average inventory.

Using *average inventory* allows us to compare inventory value at different times to measure progress against a goal.

Starting condition:

Lead time = 13 weeks   MIN = 2   EOQ = 1   Price = $1000
2 years of usage and 2 failures
Usage over LT = Failures ÷ Number of lead times = 2/8 = 0.250

Average inventory = MIN + EOQ − (Usage over LT)
              = 2 + 1 − (0.250) = 2.750 units @ $1000 = **$2,750**

Condition two years later: { LT still = 13 weeks and EOQ = 1}

2 more years of usage and 4 more failures   MIN still = 2
Usage over LT = Failures ÷ Number of lead times = 6/16 = 0.375

Average inventory = 2 + 1 − (0.375) = 2.625 units @ $1000 = **$2,625**

Planned average inventory has decreased by $125 over the two years

Figure 8.2 Comparing Inventory Over Time.

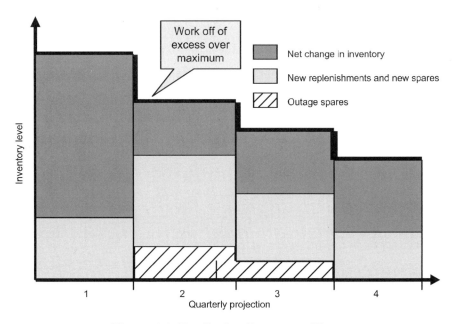

Figure 8.3 Predicting Inventory Change.

- New purchases will offset attrition as spares are used through out the year. Figure 8.3 illustrates this change in net inventory.
- Cycle count adjustments may cause errors from prior periods to increase or decrease the quantity on-hand throughout the measurement period. These adjustments are seldom predictable and need to be neutralized for measurement purposes.
- Outage and emergency spare purchases tend to be volatile; they should not be co-mingled with any goal for controlling safety stocks.
- A part may be declared obsolete during the year; it should be removed from the storeroom and taken out of the measurement.

If all of the factors above are considered, or excluded, when setting goals, it is possible to have a well-conceived goal.

8.4   DECIDING WHAT TO MEASURE

One of the first steps in goal setting is deciding what to use for performance measurement. Setting too many goals will defeat the purpose of monitoring. It is usually best to limit the number of goals to only three or four. In the following sections, we will discuss our choices for the top three goals for tracking inventory performance for active and rarely-used inventory.

8.4.1 Active Inventory

Because active inventory moves through the system fairly rapidly, over and understocks can self-correct quickly. For this reason, the measurements used to track performance must be more dynamic than those used to track slow-moving inventory. Our top 3 choices for active item metrics are service level, turnover, and order code distribution.

Service Level

We discussed service level in detail in Chapter 2. Over the years, it has been accepted as one of the best measures of performance for active items because: 1) it is a concept that is easy to understand, 2) it can be measured accurately, and 3) it can be changed quickly. For example, everybody can grasp the fact that a 95-percent service level (meaning only 5 percent of the calls at the storeroom window go unfilled) is better than a 90-percent service level.

Turnover

For active items, turnover is a well-accepted measure of performance because the inventory is moving rapidly through the storeroom. The higher the turnover ratio, the higher the return on the inventory asset. In Figure 1.3 (Chapter 1), we showed turnover as sales divided by total assets. The companion meaning of turnover at the storeroom counter is total value of withdrawals divided by total value of spares issued.

Order Code Distribution

In Chapter 10 we will discuss what we consider one of the best practices for managing active inventory: the Stock Management System (SMS) offered by Decision Associates, Inc. One of the key reasons we like SMS is the order code distribution it routinely calculates (See Figure 10.11 for details of the codes). In this system, a code of 0 would mean the stocking levels are appropriate (balanced) for meeting demand during the next lead time. Codes of -1 or +1 mean the stock levels are not optimum, but are acceptable for normal ordering during the next lead time. Therefore, any item with a code in the +/-1 range could be considered balanced for planning future procurements, whereas codes outside the +/-1 range need to have the item MIN/MAX adjusted.

8.4.2 Rarely-Used Inventory

Because rarely-used inventory moves slowly, some of the measurements of performance used for active items don't work for these items, turnover being one of the major ones. In Chapter 10 we will discuss our choice for one of the best practices for managing rarely-used inventory, the RUSL process offered by Inventory Solutions, Inc. Our top three choices for rarely-used metrics are the absolute variance ratio (AVR), the +/-1 percentage, and constant dollar change in inventory value.

The Absolute Variance Ratio (AVR)

The AVR provides a valid comparison of the relative inventory balance between companies, or among plants within the same company. Figure 8.4 lists some of the main reasons the AVR is a better measure than turnover for slow-moving inventory. Figure 8.5 shows a sample calculation for computing the AVR.

AVR compares the relative inventory balance among companies or plants within the same company.

Main reasons for using AVR instead of turnover for measuring slow-moving production spare parts:

- AVR measures the quality of the inventory plan.
- *AVR* quantifies the deviation of the plan from an optimum benchmark.
- AVR decreases and approaches zero as planned stocks are adjusted.
- Turnover only measures the part of inventory that is used .
- Ideal turnover for spares is zero (if equipment is performing well).
- Turnover is good for retail, but not for slow-moving spares.

Figure 8.4 Using AVR to Measure Inventory Performance

AVR compares the ratio of inventory deviations (high and low) to the optimum inventory level.

Example:

Total inventory deviation high   = $500,000

Total inventory deviation low   = $300,000

Optimum inventory   = $4,000,000

$$AVR = \frac{(High) + (Low)}{(Optimum)}$$

$$AVR = \frac{(\$500{,}000) + (\$300{,}000)}{(\$4{,}000{,}000)} \qquad AVR = .20$$

Figure 8.5 Calculating AVR

An important feature of the AVR is the ability to represent it graphically (see Figure 8.6). Sometimes referred to as the "clothes line" chart because of its resemblance to clothes pins on a clothes line, the chart plots AVR versus the high and low deviation of the inventory plan from optimum. The percentage of the total inventory deviation high is represented by the amount of the bar above the horizontal line, while the percentage of the total inventory deviation low is represented by the amount of the bar below the horizontal line. The further the bar is to the left (lower AVR), the more balanced the inventory plan. If the majority of the bar is above the line, the deviation from optimum is due to planned overstocks (plants B, E, F and G in Figure 8.6); if the majority is below the line, the deviation from optimum is due to planned understocks (plants A, C and D).

Two additional points about AVR:

•   It is possible to predict what the value of the AVR will be if

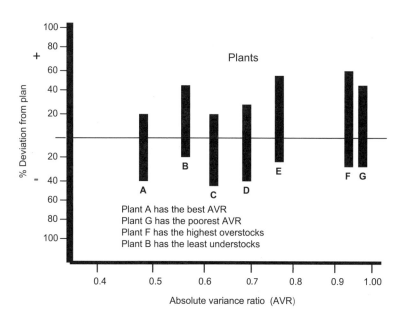

Figure 8.6 The AVR "Clothes Line" Chart

certain actions are taken to adjust inventory stocking parameters, such as changing the current reorder quantity to the EOQ amount (the what-if scenarios).

• The AVR is used to measure the balance of the inventory plan. This plan is determined by the minimum and maximum values currently used in the material management system. It has nothing to do with the balances-on-hand in the storeroom. Once minimum and maximum values are set, whether you recognize it or not, you are either trying to increase, decrease, or keep the amount of the spare the same in the future. The main objective in inventory management is to get the plan right (the MINs and MAXs set at the optimum level), and then the balanced inventory will follow.

The +/-1 Percentage

The +/-1 percentage measures the percentage of current reorder points that are set within one unit of the optimum reorder point recommendation, as determined by an appropriate decision support tool. Figure 8.7 shows a typical calculation. Because the +/-1 percentage fo-

The percentage of current order points within one unit of optimum recommended order points

Example:

Total key items	= 2,000
Current order points equal optimum	= 600
Current order points within one unit of optimum (400 one unit high; 300 one unit low)	= 700

$$+/\text{-}1\ Ratio = \frac{(Equal) + (High) + (Low)}{(Total\ key\ Items)}$$

$$+/\text{-}1\ Ratio = \frac{(600 + 400 + 300)}{(2,000)} \qquad +/\text{-}1 = 65\%$$

Figure 8.7 Applying the =+/-1Percentage

cuses on current reorder points, the value can be increased quickly by merely adjusting the current order points to more optimum levels.

Constant Dollar Change in Inventory Value

Most managers want to see the dollars and, therefore, it's usually necessary to have at least one primary or secondary goal devoted to a financial parameter. Regardless of the parameter chosen, the goal should be measured in constant dollars.

8.5   DECIDING HOW OFTEN TO MEASURE

Many factors can come into play when deciding how often a goal should be measured including: 1) whether you are monitoring a fast-moving or slow-moving event, 2) how useful more frequent feedback would be for taking corrective action, 3) whether the data base is adequately designed to capture frequent measurements, 4) the response time between making decisions on the information, and 5) how often management would really look at the results.

Our experience over the years suggests that active items should probably be monitored on a weekly or monthly basis. In that way, corrective actions can be taken quickly and deviations, such as excessive order quantities, corrected quickly. Prioritized reports will simplify the process. The need for feedback on active items tends to be required at the lower levels of the management structure.

Rarely-used inventory does not require frequent reporting to management. In most cases, quarterly reporting of results against plan is more than adequate. Quarterly reports also tend to fit in better with the demand for information from top management of the company.

8.6   WHEN INVENTORY MEASURES DON'T WORK

One of the longest standing disagreements regarding monitoring inventory is whether or not turnover is an appropriate measure in all instances. We think it is not, especially for monitoring slow-moving inventory. Although the measure may be valid for certain types of inven-

Although turnover is a valid performance measure for profitability, it is *not* a valid measure for slow-moving spare parts management.

Company	Total SKUs	SKUs "No use"	SKUs "With use"	Annual average turnover	
				SKUs "No use"	SKUs "With use"
A	28,582	17,773	10,809	0.00	0.35
B	20,760	15,394	5,366	0.00	0.81
C	31,943	26,312	5,631	0.00	0.69
D	165,027	89,901	75,126	0.00	0.86
E	13,268	7,234	6,334	0.00	0.54
F	6,659	4,533	2,126	0.00	0.67

The Maytag repairman would like these numbers.

Figure 8.8 Turnover as a Measure of Performance

tory such as transmission, distribution, or production consumables, it is a worthless measure for production plant spares for the following reasons: 1) plant turnover rates are very low because much of the inventory consists of insurance spares which, the company hopes, are never used, 2) production objectives are different among different companies, thereby making comparisons of turnover invalid, and 3) accounting practices among companies create major differences in calculating turnover rates. The latter two reasons tend to make invalid, or at best misleading, the practice of sharing data among companies to benchmark performance.

Figure 8.8 shows why. The chart shows stock movement data for six different companies. Usage history is compared over a three-year period and the SKUs sorted into "with use" and "no use". Not surprisingly, the SKUs without usage during the period have a calculated turnover of 0.00. For the items with use, the turns may be low by active inventory standards, but they are at least positive (values below 1.00 turns are normal).

Experience at hundreds of production plant around the world in different industries shows that it is normal to have 50 percent or more of production spares not move at all over the most recent three-year period. Remember, that's good news because it indicates the equipment for which the part is spared is running well; if it were failing frequently, you would be using spares.

8.7   SOME GOAL SETTING CASE STUDIES

All of the following case studies are about setting and monitoring inventory goals. In some cases, the study's purpose is to determine why performance did not meet expectations. In others, it is to compare one unit of a company with another and determine which unit performed best against goal. Somewhere among these cases is likely to be one or two that depicts a situation and a possible solution at your company.

Mechanical Wear Blocks

CASE STUDY 8-1 AJAX UTILITY

The Situation:

AJAX utility is a large southeastern utility operating both fossil-fired and nuclear generating stations. Over the past few years, the company has expanded into the non-regulated energy market by acquiring several medium-size generating stations in different parts of the country, especially the far west. In a major change from past practice, the utility hired and placed at one of their stations a new plant manager who had no previous experience in the utility industry, having in fact come from a retail organization.

Although bright, and hard working, the plant manager was somewhat misinformed about how to set goals for station performance improvement. One new goal in particular gave concern to the operations manager, who commented: "Shortly after arriving, Sandy (the new plant manager) started asking us for monthly turnover data on all our operating inventories. Although the numbers for the active and commodity stuff seemed acceptable, we were constantly told 'to get the spare parts turnover up.' To my way of thinking that was wrong, and I said so, but I was told repeatedly that at Sandy's former company, higher turnover was always good. Finally, in desperation, I asked an inventory consultant who was working with us to give me some ammunition to refute Sandy's argument. This is what the consultant came up with"

The Proposed Solution:

After discussing the situation with the operations manager, the consultant suggested they compare the last several years of plant performance using both turnover and AVR to measure inventory change. The consultant was certain that using the two different metrics would show vastly different conclusions.

The Numbers:

Figure 8.9 compares results after performing a review in 2002 and monitoring progress again in 2004. A total of 3,563 key items of rarely-used spare parts were *reviewed in detail. This 16 percent of the total station items represented $19.3 million of actual inventory in 2002, approximately 81 percent of station inventory value. Optimum station inventory for these key items was determined to be $13.7 million in 2002 when applying the desired availability level for each item. The recommended inventory level assumed that all overstocks were eliminated and all understocks were increased to the optimum level regardless of part criticality. Using the station's current reorder points and reorder quantities for each key item, the station was planning to work down inventory to about $16.6 million. Because the inventory was rarely used, however, many years would be required to reach the planned inventory goal.*

The two indicators were calculated to determine performance in 2002. The AVR of 0.62 was better than average for most utility stations reviewed. The percent of key items with reorder point within one unit, plus or minus, of the optimum reorder point was determined to be about 60 percent (about normal for utility stations). The turnover ratio for 2002 was 0.22.

Using data from 2004, these same items were reviewed again. By this time, over $3 million of excess inventory had been reduced by not replenishing overstocks and disposing of some obsolete spares. Most important, the AVR had decreased to 0.31 and the +/-1 percentage

Item	Initial 2002 review	2004 update
Items reviewed	3,563	3,563
Recommended Invy ($)	13,758,500	12,830,600
Actual inventory ($)	19,351,700	16,166,600
Planned inventory ($)	16,596,100	10,666,100
AVR	0.62	0.31
Per cent of order points within ± 1 of ROP	59.6	85.8
Turnover ratio for year	0.22	0.24

Figure 8.9 Comparing Results Over Time

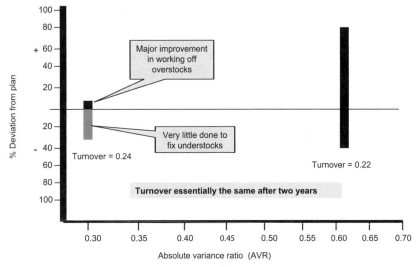

Figure 8.10 Graphing Changes in AV

increased to 85.8 percent. Figure 8.10 shows the change graphically.

A further look at Figure 8.9 reveals an interesting point about adjusting understocks. Notice that the planned inventory in 2004 was $10.6 million, or about $2 million below the suggested optimum. In other words, the station management did not plan to increase all understocks to the recommended level for two reasons: 1) they were willing to accept the risk of not doing so, and 2) they were sure they could acquire extra spares, if necessary, from another AJAX station.

Now compare the turnover values. Even though actual inventory had decreased by 16 percent, and a risk-based plan was implemented in the 2004 update to reduce inventory by another $6 million, the turnover ratio only showed a minor change from 0.22 to 0.24.

The Conclusion:

Presented with these results, the station manager agreed to drop the use of turnover and instead started using the AVR for measuring performance for the slow-moving spare parts. This change was much to the delight of the operations staff.

CASE STUDY 8-2 AJAX GAS GROUP

The Situation:

The AJAX Gas Group is a major distributor of natural gas to the east coast and midwest sections of the United States. The group operates seven major gas processing facilities in the southwest and numerous pumping stations throughout the east and midwest. Flush with cash from the run-up in natural gas prices over the last three years, the group made major capital investments to streamline facilities, cut costs, and improve the ROI. One asset-oriented goal was to lower plant inventory, especially spares, to prudent levels.

"We've made some major improvements over the last couple of years" states the Group VP, "but there's a lot more to do. I'm a numbers person and I like to measure and reward performance based on the numbers, not some subjective basis. We've done a lot to improve our inventory numbers and I just last quarter put out a new basis for monitoring it. Let me show it to you."

The Proposed Solution:

"In the past I just checked to see how my managers performed against goal. But that doesn't always tell you

AJAX Gas used the absolute variance ratio (AVR) to measure performance improvement among its divisions.

Division	AVR 12/31/03	AVR 12/31/04	Change in AVR	Rank
180	0.92	0.54	0.38	1
270	0.68	0.43	0.25	2
570	0.55	0.32	0.23	3
330	0.59	0.40	0.19	4
440	0.80	0.65	0.15	5
750	0.66	0.56	0.10	6
510	0.71	0.67	0.04	7

Figure 8.11 Using AVR to Measure Performance

who really did the best job. Now I measure who made the biggest improvement against goal."

The Numbers:

The seven group processing plants began using AVR for measuring inventory performance in 2003. Figure 8.11 shows the change in AVR from the end of 2003 to the end of 2004. If achieving the lowest AVR was the measure of performance, Division 570 would rank No. 1 and Division 510 would be seventh. When the ranking was changed to change in AVR, Division 180 became No. 1, although Division 510 still came in last.

The Conclusion:

Bonus allocations for 2004 were distributed on the basis of the improvement in AVR ranking. The Division 180 manager was happy; Division 510's manager was not.

 CASE STUDY 8-3 AJAX WORLD UTILITY

The Situation:

AJAX World Utility is the eleventh-largest generating utility in the world. All of its generation plants are fueled by coal or natural gas. The VP for Power Generation was a stickler for setting and monitoring performance against pre-set goals. The company had the usual goals for plant utilization, such as cost per kilowatt generated and safety lost-time accidents. But the VP recognized early that more specific goals would be needed to manage the total assets assigned to the Generation Division, and two years ago set goals for the first time.

The VP put it this way, "I've talked to a lot of managers in the same job I'm in at other plants. They tell me they limit what they monitor to just a couple categories of inventory like active items and new spares. I don't think that's enough if you're really going to control the entire inventory asset. So two years ago I asked my staff to put together an all-inclusive set of goals to monitor. I like what they came up with."

The Proposed Solution:

The staff recognized that inventory could be broken

AJAX World Utility used this approach to monitor inventory dollars

Component of inventory	Value on Jan 1 ($000)	Target on Dec 31 ($000)	Actual on Dec 31 ($000)
[A] Rarely used:	63,855	60,900	56,962
All rarely used	63,885	60,000	56,109
To improve availability	-	900	853
[B] Active items	12,257	9,800	11,210
[C] Inflation projection	-	3,500	2,462
[D] Anticipated new spares	-	800	1,055
Scheduled overhaul items	5,285	7,500	7,690
Reserved items	1,790	2,600	2,425
Strategic spares	9,880	10,500	10,106
Total inventory	93,067	95,600	91,910

☐ Target goal for year not met

Figure 8.12 Setting and Monitoring Stock Targets

*down into two distinct categories: dependent demand and in-
dependent demand. Dependent demand included items that
could be precisely scheduled and procured accordingly, such
as scheduled overhaul spares and strategic spares (their name
for capital spares). Independent demand was driven by generating commitments and included both active and rarely-used
spares. Values for the various categories were developed and a
set of targets for year-end were set during meetings between the
VP and the individual plant managers. Figure 8.12 shows the
goal chart for the entire Generating Division. Each plant also
had its own goal chart.*

The Numbers:
*As Figure 8.12 shows, goals for all categories were met for
the year except for active items, new spares, and scheduled
overhauls. The reasons for failing to meet goal for these categories were discussed extensively by the VP and the responsible plant manager at their regularly scheduled monthly meetings.*

*Along with the financial goals, the VP also monitored the
AVR and +/-1 percentage to track inventory balance. These
metrics are shown in Figure 8.13 and 8.14. As expected, the*

The target AVR for all of AJAX World Utilityís stations was 0.40.

Station	Qtr 1 AVR	Qtr 2 AVR	Qtr 3 AVR	Qtr 4 AVR	Target met?
A	0.83	0.62	0.46	0.34	Yes
B	0.92	0.71	0.58	0.42	No
C	0.79	0.82	0.71	0.55	No
D	0.74	0.49	0.47	0.37	Yes
E	0.80	0.49	0.39	0.29	Yes
F	0.69	0.41	0.36	0.31	Yes
G	0.55	0.40	0.41	0.38	Yes
H	0.62	0.50	0.41	0.33	Yes
I	0.72	0.60	0.42	0.37	Yes

Figure 8.13 Monitoring AVR

+/-1 percentage correlated closely with the AVR; that is, the lower the AVR, the higher the +/-1 percentage.

The Conclusion:

The goal charts used to monitor inventory had made a major contribution to the $6 million net improvement in rarely used inventory during the year. The VP, made inventory improvement a shared goal among the plant manager, the maintenance manager, and the plant administrative manager. They all got judged accordingly and collectively.

 CASE STUDY 8-4 AJAX NUCLEAR

The Situation:

AJAX Nuclear was a 985-megawatt nuclear-fueled generating station in the eastern United States. Unscheduled outages had caused the plant to shutdown, once for eight days in January, and again for fourteen days in March. During these times, the utility was required to purchase replacement power, often at prices far in excess of their own cost to generate.

The effect was putting a financial strain on the company; a directive was sent down from top management to

The target +/-1 percentage for all of AJAX World Utilityís stations was 90.0.

Station	Qtr 1 %	Qtr 2 %	Qtr 3 %	Qtr 4 %	Target met?
A	78.2	90.3	91.2	91.7	Yes
B	65.3	77.0	84.5	90.9	Yes
C	72.4	86.9	84.7	88.2	No
D	57.5	76.2	78.9	92.3	Yes
E	72.4	81.1	90.6	93.1	Yes
F	84.2	93.2	93.9	92.7	Yes
G	80.1	89.5	91.6	93.9	Yes
H	81.1	91.3	92.2	92.6	Yes
I	57.0	80.4	89.6	91.8	Yes

Figure 8.14 Monitoring +/-1 Percentage

cut all possible corners to offset the revenue loss by year-end. The station supply-chain manager, who was given an edict to cut inventory by 18 percent, reacted as follows, "I don't know who came up with that number, but they must be out-of-their-mind. Our inventory moves too slowly to just wave a magic wand and make it disappear."

The Proposed Solution:

The supply-chain manager decided to put some numbers together to prove the goal was unattainable. Two assumptions were made: 1) All current open purchase orders could be cancelled, and 2) no new purchase orders would be written unless urgent. Investment recovery of excess inventory was not an option because of the write-off cost; therefore, all of the decrease would have to come from working down existing inventory over the next eight months.

The Numbers:

The station storeroom contained 46,665 stock items with a current inventory value of $38.2 million. To meet *the 18-percent decrease, the stock would have to get down to $33.0 million. An analysis showed that only about $2.9 million of current inventory could possibly work-off by year-end, not the $5.2 million required to meet the goal. Even more alarming was the conclusion that over 3,500 stock items would be in a backorder condition by year-end if procurements were stopped.*

The Conclusion:

When presented with the numbers, the management cut the goal from 18 to 5 percent. The supply-chain manager put a poster on the office wall. It read "Don't Set Impossible Goals."

 CASE STUDY 8-5 AJAX OXYGEN

The Situation:

AJAX Oxygen is a large specialty gas company operating two processing plants, one in Alabama and one in Texas. The principal use for the oxygen was a new steel plant built adjacent to their Alabama facility, and a light-ends refining facility near their Texas plant. Because prices for oxygen had been increasing lately, the company was able to apply some of the gain to abnormal changes against the bottom-line for obsolescence and inventory write-off. Performance for the company in 2004 was good and the president wanted to show the board of directors some performance charts. "I'm very proud of what we've accomplished with our inventory program over the last year, and I want the directors to know about it."

The Proposed Solution:

The supply-chain manager was asked to prepare a brief presentation outlining the recent actions taken to improve inventory, and showing just one summary chart.

Performance measure	Alabama plant		Texas plant	
	12 months since implementation		6 months since implementation	
	Original	Current	Original	Current
Plant actual inventory ($ millions)	7.8	5.6 *	16.4	10.5 *
Plant inventory plan ($ millions)	5.8	4.0	10.5	7.1
AVR	0.70	0.39	0.78	0.37
Percent reorder points within ± 1 unit of recommended	67%	94%	66%	93%

* Includes substantial write-offs   ▢ Superior performance

Figure 8.15 Inventory Improvement Performance

The Numbers:

When implementing the new inventory initiative, it was decided to pilot the program first at the Alabama plant (even though it had less inventory). If successful, the program would then move to the Texas plant within six month (the Texas plant was still adding spares from a recent expansion). Figure 8.15 shows the summary table prepared for the presentation. Both actual inventory and planned inventory were shown, along with AVR and +/-1 values. The supply-chain manager asked the consultant working with them on inventory to advise which of their results would be typical performance (about average) and which would be considered superior performance (in the top quartile for all industries).

The Conclusion:

With the help from the inventory write-down account, the decrease in actual inventory at both plants was determined to be superior performance, as was getting the +/-1 percentage over 90 percent.

The Board was happy with the performance.

☺   ☺   ☺   ☺   ☺   ☺   ☺   ☺   ☺   ☺

CASE STUDY 8-6 AJAX DIESEL

The Situation:

AJAX Diesel is a manufacturer of medium- and heavy-weight diesel engines for the truck market. The company's only plant is in Indiana and serves both the domestic and international markets. Two years ago, the company implemented a series of cost cutting initiatives including one to adjust spare parts stocking levels. As part of the program, the consultant predicted that storeroom spares would decrease over the next year (on a constant dollar basis) by taking the actions recommended. Results after one year suggested otherwise. "Instead of a decrease in inventory I'm looking at over a half million dollar increase" said the plant manager. "What's going on?"

The Potential Solution:

The consultant agreed to look at the impact of price changes and other factors that may have caused the increase.

The Numbers:

The entire inventory at AJAX Diesel was focused on only 1,200 rarely-used key items (RUKIs). The analysis showed that 302 of the RUKIs increased by $1,160,000, whereas the rest of the items showed a decrease or stayed the same, for a net inventory increase of $717,700 (See Figure 8.16). Price increases since the project startup accounted for $140,000 of this increase. Balances on hand for 212 items were at or below the reorder point when the benchmark file was created, and procurement or other actions increased the benchmark inventory by $530,200. Several possible actions may have caused the increase:

- *Many of the items were purchased before the program startup and arrived afterwards.*
- *Some of the items were returned to the storeroom by maintenance.*
- *The balances for some items increased because of cycle count adjustments.*

Item	Inventory impact ($)
Total net inventory increase from beginning (benchmark) file	717,700
Net price increases for 1,200 RUKIs since benchmark file	140,100
Constant dollar increase since benchmark file	577,600
Inventory added after project start from actions before benchmark file	530,200
Adjusted constant dollar increase since benchmark file	47,400
Inventory value of items with BOH in latest file greater than plant MAX	150,700

Figure 8.16 Analysis of Inventory Files

- *Items were returned to stock from overhauls or capital projects.*
- *Data entry errors caused the balances to be artificially high.*

As shown in Figure 8.16, if the $530,200 increase was considered outside the window of the project, the net constant dollar increase would have been only $47,400. However, $150,700 of increases were from actions that caused the balance-on-hand of certain items to be over the maximum stocking level specified in the material system.

The Conclusion:
Deducting the "over MAX" items ($150,700) from the constant dollar increase resulted in an actual net decrease of $103,300 for the year.

The plant manger response was "Why are we adding inventory over our own maximum limits"? Good Question.

CASE STUDY 8-7 AJAX VANADIUM

The Situation:

AJAX Vanadium is the second-largest producer of vanadium in the world. It owns mines and processing facilities in Chad and Chile, where Carnotite ore is processed into vanadium oxide, and in Australia and Indonesia, where Vanadinite is processed to oxide. Smelters in Aruba and Australia transform the vanadium oxide into ingots for use by customers for rust resistant and high-speed tool steels.

Last year the general manager of the Aruba smelter instituted an inventory control program to project changes in plant inventory. Estimates of excess inventory work-off, new spares purchases, and overhaul spares were made for the year in an attempt to forecast the change in total spares inventory throughout the year. Figure 8.17 shows a quarter-by-quarter summary of the results and the deviation of actual inventory levels from projected levels. The general manager asked, "Why the wild swings from quarter-to-quarter? Are we missing something?"

The Proposed Solution:

The plant's supply-chain manager, who suspected they were in fact missing something in setting their projections, set out to look at the data in more detail, especially within the procurement process.

The Numbers:

The estimates for each line item category in Figure 8.17 were reviewed in detail. The quarterly projected estimates for work-off of excess inventory, new spares purchases, and projected overhaul spares purchases proved to be reasonably close to estimate. However, sizable deviations were observed in the quarterly estimates for normal replenishment of existing stock. The supply-chain manager had a hunch he know what the problem was and contacted the supply-chain manager at the Australia smelter, requesting some data from that plant. Figure 8.18 shows what was

Item	Inventory change from benchmark through quarter shown ($)			
	First	Second	Third	Fourth
Projected excess inventory work off	-80,000	-78,000	-75,000	-70,000
Projected existing spares replenishment	+350,000	+500,000	+500,000	+450,000
Planned new spares	+112,000	+60,000	+60,000	+75,000
Projected overhaul purchases	+62,000	+81,000	+67,000	+61,000
Actual price increases	+84,700	+151,600	+156,000	+149,300
Projected change	+528,700	+714,600	+708,000	+735,300
Actual change	+585,300	+999,800	+962,600	+762,500
Deviation (Actual minus projected/actual)	+9.7 %	+ 28.5 %	+26.4 %	+3.6 %

Figure 8.17 Inventory Cash Flow Analysis

The number of items awaiting replenishment can vary greatly

Location	Key items awaiting replenishment at program start (%)
Aruba	23.5
Australia	6.7

Figure 8.18 Queuing for Replenishment

discovered. The Aruba plant had nearly four times as many spares awaiting procurement when the benchmark inventory level was established compared to the Australia plant.

The Conclusion:

The estimates for replenishing existing spares failed to consider the higher than normal level of items in the purchasing queue, as well as the longer lead times to replenish many of the items. The main reasons for the higher queue were delays in placing purchase orders, awaiting updated vendor price quotes, and a shortage of staff due to vacations and sickness.

CHAPTER 9

PARTNERING WITH OTHERS

9.1 WHAT THE READER SHOULD LEARN FROM THIS CHAPTER:

- How partnering can be an advantage
- How you can certify Availability
- An overview of firms that can help optimize your inventory
- Case studies related to partnering

9.2 WHY CONSIDER PARTNERING

Companies have two choices when they consider implementing an inventory initiative. They can go it alone or they can get help from others. Over the years we have seen companies go both ways, with the "get help from others" option usually producing better, faster results.

The logic used by most companies that decide to go it alone is that they have the talent and resources in-house to fix their problems. Almost always they have skilled financial and accounting people, supply-chain people, and operating and maintenance people who understand the problem. But what they frequently don't have is some of the specialized analytical competence and skills needed to solve the problem efficiently. For example, in Chapter 5 (Case Study 5-2) we showed how one company spent over one year reviewing the wrong items because they failed to determine in advance which were key items. We have seen numerous other situations in which the company ultimately got to a good solution, but spent far more time and resources getting there than would have been required if they sought outside help.

Using outside help to implement an inventory initiative can be beneficial for a number of reasons: 1) the outside party may have unique competencies that are not available in-house, 2) they usually have solved similar problems before and know how to limit the use of internal resources, 3) they can bring some of the best practices used by other companies to solve the problem, and 4) they can often overcome internal conflicts by bringing a fresh face to the discussion. Offsetting these benefits are some common internal concerns that arise when consultants or other contract personnel are hired: 1) in-house managers often feel their ability may be questioned if gross overstocks or inventory imbalances are found, 2) outsiders really can't appreciate our problem because they don't understand our internal politics, and 3) they will simply offer a quick fix, take our money, and fade away. All of these are perfectly valid concerns.

One situation we have seen, especially over the last five years, has deferred many inventory initiatives, regardless of whether the program was done entirely with in-house resources or with outside help; this is the sharp cutback in staff at most companies during the recent recession. We have seen purchasing and stores staffs cut in half, with those left standing expected to cover the previous work load. Likewise, operating and maintenance resources have been curtailed, making it difficult to get any of their time to support an inventory initiative. Obviously, financial resources have been cut as well, especially for discretionary projects, of which inventory is a prime example.

In the remaining sections of this chapter we will discuss some of the outside firms that can play a role in helping companies optimize their inventory asset. Much of the information shown was collected from Internet web sites and from discussions with the firms. We also will offer some interesting case studies.

9.3 POTENTIAL PARTNERS

In the following sections, we will discuss partnering with various firms. We believe that the firms listed, in the market niche they serve, can provide a direct benefit that will contribute to optimizing the inventory asset.

9.3.1 Partnering With Distributors

Buying spare parts from a distributor rather than stocking in-house can save money. It can also be more efficient.

In his book *Competitive Strategy*[*], Michael Porter states that the success or failure of any firm depends on competitive advantage-delivering the product or service at lower cost or offering to the buyer unique benefits that justify a premium price.

Many distributors seek to gain a cost advantage over competitors by doing some of the following:

- Leveraging the cumulative total value of the savings derived from negotiating lower unit cost.
- Using multi-model forecasting software tools to improve service levels.
- Lowering transaction costs by implementing programs such as integrated supply chains, vendor alliances, consignment, and the use of credit cards.

But many of the above techniques fail to achieve optimum stock investment for both the distributor and the customer. In many cases, the end result is only the movement of inventory from the customer's storeroom to the distributor's warehouse. Although the lower inventory asset level reflected in the customer's annual report looks impressive, we all know there is no free lunch.

One technique gaining acceptance is to certify availability of non-forecastable spares by the distributor. Under this concept, both the customer and the distributor can benefit from having collectively optimum (usually lower) stocking levels, while at the same time ensuring spare part availability when needed. An outline of the concept developed by Inventory Solutions, Inc., of Akron, Ohio follows:

- The distributor determines which rarely-used parts to certify to the customer.

---

* Porter, Michael E., *Competitive Strategy*, The Free Press, (1985).

- The customer decides whether they need same day, same week, or two-week delivery for each spare part.
- The distributor contracts with the consultant to supply the recommended reorder point and economic order quantities to meet the customer's delivery requirements for all spare parts.
- The customer stocks the reorder point amount (MIN) at their facility and the distributor stocks (or reserves) the EOQ amount at their warehouse.
- The distributor evaluates its total demand for the certified items in order to meet the demand from all customers for the items.

The cost benefits of the Certified Availability concept or shown in the following case study.

Changes to plant stocking	Changes to supplier stocking
50% of items were set to lower MIN, saving $16,885	18% of items were set to lower EOQ, saving $10,119
31% of items were set to higher Min, adding $3,057	61% of items were set to higher EOQ, adding $9,918
19% of items had no MIN change	21% of items had no EOQ change
Net plant overall savings were $13,748	Net supplier overall savings were $201

Both the plant and the supplier achieved a better balanced inventory.
Both eliminated overstocks and understocks.
The plant saved the most, but the supplier gained a *competitive advantage*.

Figure 9.1 Partnering With Your Distributor

CASE STUDY 9-1: AJAX METALS

The Situation:

AJAX Metals operates a metal extrusion and finishing plant in the eastern United States. Recently a consultant contacted them and suggested they consider trying a new inventory sharing concept called Certified Availability with one of their local distributors. "We're always looking for ways to cut our inventory carrying costs and this concept proposed by the consultant sounded pretty good. We decided to give it a try."

The Proposed Solution:

A meeting was held with AJAX Metals and one of their key distributors to determine which spares to include in the study and how best to collect the necessary input data needed to determine recommended stocking levels.

The Numbers:

Eighty-two rarely-used items currently under a blanket order agreement and ranging in price from fifty cents to $3,434 were selected for analysis. AJAX Metals determined for each item whether they needed same day, same week, or two-week delivery from the distributor. Data files containing stock descriptions, price, usage demand, current MIN/EOQ stocking levels, and balances-on-hand were supplied to the consultant. Figure 9.1 summarizes the relevant results.

The Conclusion:

From the analysis, AJAX Metals benefited the most inventory savings ($13,748 versus only $201 for the distributor), but the alliance between the two parties was strengthened, giving a competitive advantage to the distributor in future dealings with AJAX.

9.3.2 Partnering With Your Original Equipment Supplier (OEM)

You probably bought a lot of equipment from the OEM in the past, so it makes sense to work with them in the future to supply replacement spares as long as they give you competitive prices and deliveries. Sooner or later, there will always be other choices on where to buy replacement spares, but those choices can carry substantial risk. Specifications on spares can often change over time. Having someone (often called pirates) make replacement spares to old drawings can result in buying a spare that doesn't work for the intended purpose. Frustrating pirates is one reason many OEMs regularly change stock numbers and make subtle changes to the design drawings for their equipment parts.

No one set of objectives and tools works best when buying both active and slow-moving spares. (See Figure 9.2 for an overview.) Therefore, partnering with your OEMs to meet delivery of spares makes as much sense as it does for working with your distributors. In fact, many companies offer what they call Assured Stocking programs that guarantee delivery at fixed prices to customers who sign up for the program. Figure 9.3 shows how a typical program is structured. The last point noted in the figure-updating stock levels to support all cus-

---

No one set of objectives and tools works for both active and slow-moving items.
For active items, the objective is to improve service level, reduce item cost, and minimize transportation costs.
Active items can be forecasted and backorders avoided.
Slow-moving items (rarely-used items) can not be forecasted. These three statements can be made:

- You have no idea when a rarely-used part will be needed.
- No amount of inventory can avoid a backorder.
- No amount of inventory reduction can offset for the lack of a needed critical part.

For slow-moving items, the objective is to minimize total risk cost.

Figure 9.2 Partnering With Your OEM

tomers is important because the leverage of larger demand can convert a rarely-used spare into an active spare and allow forecasting techniques to set more accurate stocking levels. The following case study between AJAX Generation and one of its OEMs illustrates the assured stocking concept.

A supplier maintains a stock of inventory to reduce the risk of customer downtime during production:

- The supplier retains ownership of inventory until withdrawn.
- Inventory forecasts and usage data are exchanged between the supplier and the manufacturer.
- A planned stocking level is determined using algorithms for either active or slow-moving items.
- Stocking levels are updated periodically as more usage data become available.
- The supplier updates stocking levels to support all customers.

Figure 9.3 Assured Stocking Program

Part XYZ costs $1,000 and takes 4 weeks for normal delivery from the OEM to the station.

⟶ Required amounts ⟵

Case	Yrs of usage	Total issues	Lead time (Weeks)	MIN	MAX	EOQ	Maximum amount station carries (Units)	Amount OEM carries (Units)
A	3	2	4	1	2	1	2	0
B	3	2	0.14 (1 Day)	0	1	1	0	1
C	3	24	4	3	6	3	0	6 (Reserves 1 unit for AJAX)

CASE A: The station stocks Part XYZ and buys directly from the OEM.
CASE B: The station buys from the OEM under an Assured Stocking Program.
CASE C: The OEM stocks enough parts to meet requirements for all of its customers.

Figure 9.4 Partnering With Your OEM

CASE STUDY 9-2: AJAX GENERATION

The Situation:

AJAX Generation operates a 640-megawatt coal-fired power station in Kentucky. Recently one of its OEMs proposed to them that they consider an assured stocking program for selected spares that have had demand of more than one unit over the last three years. The station supply-chain manager's answer was, "Sure, give me some details on what you have in mind and let's see if it makes sense."

The Proposed Solution:

The OEM proposed that they study three cases: 1) the current stocking situation at the station, 2) the stocking required at the station under a one-day assured delivery program, and 3) the impact on the OEM if all of their customers for the parts used assured stocking.

The Numbers:

Figure 9.4 shows the results of the study for one item (Part XYZ). Under current practice (Case A), the plant stocks part XYZ to a MIN of 1 unit and a MAX of 2, buying one unit at replenishment time, which normally takes four weeks. After the one-day delivery was factored into the stocking calculation, the MIN/MAX dropped to 0/1 (Case B). When the demand for the part from all customers was considered, the OEM needed to carry six units in their warehouse to meet the total demand (Case C).

The Conclusion:

AJAX agreed to sign up for the assured stocking program for 23 spares. In each case, for a small price premium, the OEM agreed to reserve one unit of each part in their warehouse for the exclusive use of AJAX.

9.3.3   Partnering With Stocking Consultants

Regardless of how well a company manages the procurement of its spare parts, all becomes irrelevant if it doesn't set appropriate stocking levels in the material management system. We showed earlier in Chapters 1, 3, and 4 how differences between active and rarely-used inventory can affect stocking considerations when setting MIN/MAX levels. Whether stocking levels are set using forecasting techniques (active items) or risk-base techniques (for slow-moving items), the process is not simple. Most companies would be better off getting help from others with experience when trying to do this, especially if staff levels are at all-time lows.

Because the setting of stocking levels for production spare parts tends to be a niche market, only a few consulting firms in this arena have both the tools and experience to do the job efficiently. Figure 9.5 lists two of the best-known firms who have been around long enough to know this business. Each has chosen its own niche: 1) ISI exclusively on rarely-used inventory, and 2) DAI mainly on active inventory.

Both have designed their software to focus on specific problems: 1) ISI on setting MIN/MAX levels for new and replacement spares, investment recovery of excess inventory, and preventing unnecessary purchases, and 2) DAI on setting optimum stocking levels, providing comprehensive forecasting, and determining replenishments. Both offer some type of marketing incentive to induce potential customers: 1) ISI a no-cost Minimum Benefits Analysis, and 2) DAI a demo of the software using potential client data.

The following case study illustrates the use of DAI's order codes.

CASE STUDY 9-3: AJAX DISTRIBUTORS

The Situation:

AJAX Distributors is a subsidiary of AJAX Utility. It operates and maintains 172 sub-stations in the Indianapolis areas as well as servicing power lines to residences served by the sub-stations. Lately the number of backorders and stock-out for many of their SKUs has increased to levels that were beginning to cause complaints from customers based on AJAX's slow response to outages and service line upgrades.

The Proposed Solution:

The decision was made to buy an improved software package. This package incorporated several forecasting models to set stock levels for active items. One unique feature of the software was the classification of each SKU into one of seven order codes that equated to inventory balance.

The Numbers:

Figure 9.6 shows the change in order code distribution before and after running the inventory through the software and adjusting reorder points for most of the items. Notice the significant improvement in the number of balanced items from 63 to 90 percent. Also important was the drop in the -3 and -2 codes, cutting the backorder and stock-outs from 24 to 6 percent.

The Conclusion:

The decision was made to re-run the SKUs though the software weekly and make adjustments as appropriate. A goal of 90+ percent balanced was also set.

Category	ISI (RUSL)	DAI (SMS)
Company name	Inventory Solutions Inc.	Decision Associates Inc.
Principal offices	USA /UK	USA /UK
Website address (www)	Inventorysolutionsinc.com	Daiglobal.com
Software acronym	Rarely Used Inventory Stocking Logic (RUSL)	Stock Management System (SMS)
Type of part activity handled	Exclusively slow-movers (<12 Issues per year)	Mainly active Items, but defaults no use slow- movers
Software design focus	Set MIN/MAX levels; Batch or single item processing	Set MIN/MAX levels with forecasting models for active items; Mainly batch processing
Years in use since	1983 (Over 700 sites)	Early 1980s
Software features	Handles replenishments, new spares, investment recovery analyses, sensitivity analyses	Handles BOM, DRP, finished goods, and spares inventories using ordering codes; Sets safety stock using forecast error and service level
Access to software	Via internet	User hard drive or network

Figure 9.5 ROP/ROQ Consultants....continued on next page

Category	ISI (RUSL)	DAI (SMS)
Report features	MIN/MAX priority reports, top 10 reports, order quantity by EOQ, what ifs	Forecast analysis, MIN/MAX plan with 7 order codes, performance summaries
Material system link	Stand-alone decision tool, results can be exported to management system	Can be linked to stock status system to automate ordering process
License required	Yes	Yes
Client base	All industries worldwide ; About one-half utilities	All industries worldwide; Mix of all industries
Additional consulting services available	Shared spares, investment recovery, lead time Analysis	Mainly software support
Implementation approach	Client provides database, client sets criticalities and reviews/approves results	Client provides database; Client can adjust or modify recommendations
Training included	Yes	Yes
Commercial aspects	Free benefits analysis; Pay-After-You-Save (PAYS)	Offers free analysis of client database and free SMS trial
• Claim-to-fame†	No client backorders for critical items set with RUSL	Seamless management of entire inventory on PC

(The content of this figure has been reviewed by ISI and DAI.)

Figure 9.5 ROP/ROQ Consultants (Con't)

Order codes

	- 3	- 2	- 1	0	+ 1	+ 2	+3
Before	7 %	17 %	1 %	55 %	7 %	1 %	11 %
After	2 %	4 %	9 %	71 %	10 %	1 %	3 %

-3   Backorder
-2   Out of stock
-1, 0, +1   Balanced
+2   Overstocked with orders outstanding
+3   Surplus

Figure 9.6 Improving the Order Code Distribution

9.3.4   Partnering With Cataloging Consultants

Our work in inventory management has shown us time and again that the nomenclature used for naming and identifying spares is highly unstructured. From their earliest inception, many inventory management systems have grown up using ad hoc approaches for naming and numbering the spares. In the rush to buy and get the item in the storeroom and on the material system, part numbers and descriptions have been set without regard to possible duplication with already-existing parts. As mentioned earlier, OEMs contribute to this confusion by occasionally changing the name and number of their parts.

Regardless of how clever they may appear, computers are not smart enough to catch simple data entry errors such as entering "/" in a description "\" is needed instead. (If you think otherwise, try misspelling one character of an e-mail address and watch how quickly it either gets rejected by the server or sent to the wrong person!) Because of this limitation, material management systems often contain many variations of numbering for the same part.

Enter the catalogers. These are firms offering a broad range of cataloging services ranging from identifying potential duplications to

Category	SparesFinder	IHS-Intermat
Company name	SparesFinder	IHS - Intermat Solutions
Principal offices	UK	USA
Website address (WWW)	*sparesFinder.net*	*Intermat.com* and *ihs.com*
Type of business activity	Cleaning of materials databases for ERP systems; Data entry system to control data quality in material master.	MRO parts optimization, Includes catalog standardization, classification, duplicate ID/elimination, and content enhancement services and software
Years in business	7	27
Client base	Many large global industrial and manufacturing companies	Utilities, telecom, oil & gas, chemical, pulp & paper, pharmaceuticals, metals/mining, food & beverage, aerospace, automotive, facility management
License required	Subscription	Contract
Software features	Aggregation of disparate databases Tools to drive and manage workflow Matching to supplier catalogues Parsing nouns/modifiers and part numbers Eliminating duplicates Many users across the web to reduce costs Project coordination and control Integration with ERP/CMMS	Uses standard modifier dictionary (SMD) Uses *Struxure 5* catalog authoring/applications Can store images/graphics
Software compatibility	Compatible with most enterprise systems, e.g., SAP	Compatible with Indus, Oracle, SAP, and others

Figure 9.7 Cataloging Services

(The content of this figure has been reviewed by SparesFinder and IHS-Intermat.)

standardizing and enhancing spares on an item-by-item basis. Figure 9.7 lists one UK and one U.S. firm well known for their cataloging services. Although the approach to cataloging by each may differ, the mission of both firms is to help clients eliminate spare part duplication and avoid unnecessary replacement of duplicates. Consider the following case study.

CASE STUDY 9-4: AJAX ELECTRONICS

The Situation:
AJAX Electronics had grown rapidly as a supplier of electronic parts and components to the general industry. Unfortunately, the rapid growth caused a sizable amount of its warehouse stocks to be identified "on the fly," with little or no attempt at description standardization. Hence, the legacy material management system was of poor quality and thought to be highly inaccurate.

The Proposed Solution:
The decision was made to employ a firm to catalog about 14,000 SKUs that were suspected of having a large number of duplicates.

The Numbers:
Once some obsolete items were eliminated, a total of 13,416 SKUs were cataloged by the firm using their proprietary process. Figure 9.8 summarizes the finding after the project was completed. Surprisingly, the number of duplicates discovered was only 472 items. The vast majority of the duplication resulted from either inaccurate and improper modifiers or description characteristics. A large number of data entry errors were corrected.

The Conclusion:
The clean-up of the parts catalog was considered a success by the management. Some additional training was implemented for the storeroom staff to cut down on the number of data entry errors.

Total SKUs cataloged   13,416
Total duplicates found   472

Reason for duplication

Inconsistency in naming master noun	10 %
Improper modifier	45 %
Improper characteristics description	33 %
Data entry errors	12 %

Figure 9.8 Cataloging Can Eliminate Duplicates

Partnering With Parts Finding/Disposal Firms

We all occasionally like to get a bargain. Sometimes a company can procure a needed spare quickly and at a significantly lower price than from the OEM. Many times the replacement need not be new because the used part will work perfectly well for the intended application. Sometimes the delivery can be short enough to cut the cost of lost production due to a backorder situation.

Figure 9.9 lists a UK and a U.S. firm who are well known and specialize in finding needed spare parts. Both operate on a membership or subscription basis where data on spares availability are shared between members. Both allow access to their software by the Internet. Besides helping clients find needed spares, the services also provide an effective way to dispose of excess inventory by listing it on the network. Each has focused on supplying their service to selected markets: 1) Rapidpartsmart to utilities (currently every U.S. nuclear utility uses their service), and 2) SparesFinder to large industrial customers. Some of the benefits of using these services include: 1) reduced inventory levels, 2) lower purchasing costs, 3) access to a variety of sources for spares, and 4) reduced downtime from stock outs. The case study below illustrates one application of the service.

Category	Rapidpartsmart	sparesFinder
Company name	Scientech Inc.	sparesFinder
Principal offices	USA	UK
Website address (WWW)	Rapidpartsmart.com	sparesFinder.Net
Software acronym	Readily Accessible Parts Inventory Database (RAPID)	VPI
Type of business activity	Spares locating using on-line database investment recovery service; Replacement part solutions; Engineered parts market exchange	Spares locating using on-line search engine; Reporting on inventory situation across multiple locations;Tools to optimize spares holdings and reduce inventory held
Software features	Online listing of subscriber parts, advertising, E-commerce over the Internet, store front for memberís urgent needs;. Advanced search engine enables search by part description, SKU, part number, or manufacturer; Enter search criteria, identify surplus available, and contact information for buying/selling; Even locate equivalent replacement parts	Inventory roll-up at corporate level; Reports which show how to release value; Search engine to pool and find spares
Years in business	15 (over 500 subscribing plants)	7
Access to software	Online using gateway to gather client content and plug-ins to provide integration with CMMS and enterprise systems	Online using web browser
License required	Membership	Subscription
Client base	Power generation: nuclear, coal, oil, hydro-T&D, natural gas T&D and parts suppliers	Many large industrial global customers

Figure 9.9 Part Finding/Disposal Services

Category	Rapidpartsmart	SparesFinder
Additional consulting services available	Sourcing specialist to locate, buy, and transport parts to clients	Offers a range of consulting services upon request
Implementation approach	Access client data to retrieve information parts available through on-line catalog and special search engines; Clients can compare inventory lists to others	Provides forum for parts location; Assists in making buy/sell contacts
Training included	Customer service professionals to assist members	Offers customer support throughout implementation
Commercial aspects	Members and/or suppliers negotiate own buy-Sell Deals; No 3rd party commissions	Subscribers and/or suppliers negotiate own buy-sell deals
"Claim-to-fame"	Entire North American nuclear utilities are members	Most sophisticated technology for optimizing holdings of spares

Figure 9.9 Part Finding/Disposal Services (Con't)

(The content of this figure has been reviewed by Rapidpartsmart and SparesFinder.)

CASE STUDY 9-5: AJAX NUCLEAR

The Situation:

AJAX Nuclear operates a two-unit nuclear station in the eastern United States. For two years, they have been a member in a spare-parts sharing arrangement with other utilities that use the same reactor design. During a recent scheduled outage, they were two days from restart when a critical condensate re-circulating pump failed while being retested after a maintenance procedure. A replacement from the OEM was quoted at a two-week lead time, even if expedited through production.

The Proposed Solution:

A search of the shared spares database indicated that a duplicate pump might be available form another nuclear utility on the west coast.

The Numbers:

After checking the pump specifications and quality control certification, it was determined that the pump on the west coast was acceptable for service at the AJAX Nuclear station. Arrangements were quickly made to overnight air freight the pump to the local airport were it was picked up, installed, and tested with about six hours to spare before the scheduled restart.

The Conclusion:

Although the premium for special air freighting the pump was high, it was a small fraction of the daily cost of delaying the startup of the unit.

9.3.6   Partnering With Consignors

Earlier in Chapter 5, Section 5.5.5, we discussed consigning as one way to reduce excess inventory. We won't repeat ourselves here.

Because of the large number of consignors serving the industry, we have not tried to single out any in this chapter. However, the following case study shows a typical arrangement between a customer and consignor.

The program:
- Stock remains in customer storeroom
- Ownership transferred to consigner; Consigner stations one employee in customer storeroom
- Pay upon draw
- List of consigned items reviewed semi-annually and adjusted accordingly

Group number	Activity level (Issues/year)	Number of parts consigned	Average price of group ($)	Total group price ($)
1	< 3	None	-	-
2	3 – 10	54	168	9,072
3	11 – 50	106	372	39,432
4	51 – 100	214	412	88,168
5	101 – 1000	196	67	13,132
6	> 1000	81	121	9,801

Total $ 159,605

Figure 9.10 Consignment of Inventory

CASE STUDY 9-6: AJAX MANUFACTURING

The Situation:

AJAX Manufacturing operates a large facility in the midwestern United States where they ship replacement spare parts and assembled components to chemical, steel, and refining clients. Lately, they have been looking for ways to cut inventory costs and a proposal from a consignor caught their attention. "This looks like it could be a good deal for both of us," said the supply-chain manager.

The Proposed Solution:

After reviewing AJAX's list of 17,614 SKUs, the consignor proposed taking control of 651 SKUs with at least three issues per year, on average, as shown in Figure 9.10.

The Numbers:

Under the consignment arrangement, ownership of $159,605 of inventory was transferred from the customer to the consignor even thought the spares would physically remain in the customer's storeroom. The consignor would invoice weekly for all parts drawn that week using draw-receipts signed by the consignor's employee stationed full time in the storeroom.

The Conclusion:

Both parties benefited by the arrangement: the customer by getting $160,000 of inventory off their books, and the consignor by having a captive customer.

Category	Oracle Inventory Optimization	PeopleSoft Inventory
Company name	Oracle	PeopleSoft
Principal offices	Worldwide	Worldwide
Website address (WWW)	Oracle.com	Peoplesoft.com (You get (Oracle/Peoplesoft)
Parts activity handled	Primarily fast-movers (Active)	Primarily fast-movers (Active)
Product features/modules	Numerous modules iIncluding: • Demand planning • Inventory optimization • Manufacturing scheduling	Modules include: • Inventory • Inventory policy planning • Supply chain warehouse

(Note: Oracle acquired Peoplesoft in 2005 and they are now one company.)
(This summary is limited to materials management modules only)

Category	Oracle Inventory Optimization	PeopleSoft Inventory
Unique features	• Advanced optimization manages variability and risk	• Supports slow-moving and fast-moving item evaluation • Improves inventory performance by simulation capabilities
Inventory module features	• Advanced inventory optimization • Time phased inventory planning • Postponement strategies • Flexible product planning goals • Uncertainty/variability analysis • Handles different probability distributions • Graphical output of plans • Improved planning by simulation Inventory module features	• Performance analysis reports • Controls/tracks inventory by business unit • Statistical tracking of svc level • Automatic data collection • Bar code generation • Automatic stocking generation (ROP/EOQ) • Multiple physical activity Features (e.g., count by ABC class) • Cost and valuation management • Enables intra-unit materials movements

(The content of this figure has been reviewed by Oracle and Peoplesoft.)
(This summary is limited to material management modules only)

Figure 9.11 Enterprise Services

9.3.7   Partnering With Enterprise Software Providers

The mission of enterprise software providers (ESPs) is to provide a package of modules designed to handle the various data and reports needed to manage a business, such as sales and marketing, accounts receivable, payroll, and supply-chain management. Most break their offering into modules that can be brought on-line according to a planned schedule. Typically, several years are needed to install a complete enterprise system. Substantial in-house and consulting assistance are needed to complete the implementation. The cost of implementing such a system can be significant.

Category	Indus Asset Suite
Company name	Indus International
Principal offices	Worldwide
Website address (WWW)	Indus.Com
Parts activity handled	Primarily fast-movers (Active)
Product features/modules	Numerous modules including: • Asset management • Materials and procurement • Supply chain management
Unique features	• Allows different standards and procedures for repair orders
Inventory module features	• Monitors ROPs, safely sock and lead time requirements • User defined catalog • Warehouse configuration control • Full material receiving function • Full shipping activity function • Monitors turns and stock-outs • Supports direct purchases from vendors by the internet

(This summary is limited to material management modules only)
(The content of this figure has been reviewed by Indus)

Figure 9.12 Enterprise Services

The number of ESPs available to choose from is shrinking. For example, Peoplesoft acquired J.D. Edwards, and recently Oracle acquired Peoplesoft, although they plan to continue to market each system separately for now.

In this section we will focus on the materials management module offered by several firms. Figures 9.11 and 9.12 list some of the key features of the inventory module summarized from the web sites of the ESPs. All offer a variety of solutions to typical inventory management problems such as scheduling, procurement, demand planning, and setting of stocking levels primarily for fast-moving parts using multiple forecasting models. In most cases, standard reports are offered which can be modified to accommodate client special requirements.

Most companies have some type of inventory management system, albeit old and possibly home-grown. As the demands for managing a business require more, better, and real-time availability of data and analysis, most companies will need to upgrade and streamline their existing systems. They would probably be better served by choosing what can be purchased rather than trying to develop it.

Mechanical elbows

CHAPTER 10

BEST PRACTICES AND LESSONS LEARNED

10.1. WHAT THE READER SHOULD LEARN FROM THIS CHAPTER:

- A best practice for slow-moving items
- A best practice for active items
- A best practice for implementation
- Important findings from twenty years of optimizing MRO inventory
- Implementation issues
- Management issues

10.2 HOW WE SELECTED BEST PRACTICE COMPANIES

We selected Inventory Solutions, Inc. (ISI) as an example of a best practice for optimizing slow-moving inventory and Decision Associates, Inc. (DAI) as among the best in class for active inventory. Our selection was based on the following criteria:

- Each consulting firm has been in business for at least twenty years.
- Each has pioneered in establishing innovative technology for controlling the stocking of MRO inventory.
- Each firm has developed versatile tools that have saved companies millions of dollars.

- Each firm has streamlined its practices to decrease the client resources needed for efficient implementation.
- Each firm has maintained a strong presence over the years in its area of expertise.

We selected the Kansas City, KS Board of Public Utilities (BPU) as the best implementer for the optimization of slow-moving inventory. Besides implementing every recommendation from ISI, BPU also achieved the best improvement in inventory performance of any client of ISI in any industry, world-wide. Some of the reasons for their excellent performance will be discussed shortly.

10.2.1 Best Practices For Rarely Used Inventory

The ISI Process Overview

Figure 10.1 outlines the overall approach used by ISI to manage a client project. To provide clients with a financial basis for contracting, the company begins by performing a no-cost Minimum Benefits Analysis, or MBA. The MBA provides the client with the following: 1) a pro-

---

Inventory Solutions, Inc. (ISI), Akron, Ohio, USA

- First performs a no-cost minimum benefits analysis (MBA)
- MBA estimates overstocks, quantifying purchases that may be unnecessary over the next two years
- Project fees are fixed, based on the amount of rarely-used inventory in clientís storeroom
- Projects normally 12-18 months; can be extended yearly for an additional fee (other services available too for a fee)
- Data to ISI and results to the client are transmitted electronically
- Clients get access to ISI's WebRUSL© Internet software for running analyses of spares 24/7
- Clients get item specific recommendations for adjusting overstocks and understocks, as well as updated performance measurements
- Some on-site training is provided
- ISI experience from performing projects at over 700 plants worldwide
- ISI algorithms for rarely-used items are industry best

Figure 10.1 A Best Practice for Rarely-Used Items

file of the client inventory showing a breakdown into active and rarely-used inventory, 2) a recommendation regarding the number of rarely used key items (RUKIs) that should be analyzed in detail, 3) a usage profile of the inventory showing the number of items that had some usage, no usage, and negative usage, 4) an estimate of overstocked inventory value, 5) a projection of the value of unnecessary purchases that are likely to be made over the next two years if reorder points are not adjusted, and 6) a set of specific recommendations for improving the inventory asset.

The project begins once the client provides a purchase order or letter of intent. Projects normally last from one-to-two years, with specific benchmarks (quarterly or semi-annually) for reporting results and performance against goals. Fees for projects are fixed and are based on the amount of rarely-used inventory to be analyzed. Client data files and ISI reports are transmitted electronically.

ISI's most recent innovation is its RUSL decision support tool (Rarely-Used inventory Stocking Logic) which allows rapid transmission of data both ways in batch or one item-at-a-time for new spares and replenishment spares. Internet transmission of data provides 24/7 access. Reports to clients are both summary and item-specific in nature; they focus on the most important changes to inventory parameters, up or down, and are always sorted in priority order (Pareto at work here).

Processing The Data By The Internet

A graphic flow sheet for the WebRUSL process is shown in Figure 10.2. The process starts by the client building an input table of data in standard Excel format using a template provided by ISI. Once the data file is ready to transmit, the client logs onto WebRUSL using a password, and browses their system for the data file to upload. Upon receipt of the file, ISI sends an acknowledgment, processes the data, and transmits the recommended stocking parameters back to the client. The entire process is fast with up to 50,000 line items analyzed in the time it takes to drink a cup of coffee.

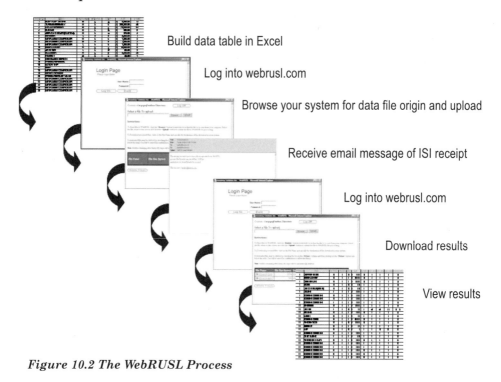

Build data table in Excel

Log into webrusl.com

Browse your system for data file origin and upload

Receive email message of ISI receipt

Log into webrusl.com

Download results

View results

Figure 10.2 The WebRUSL Process

Key Input Data

As mentioned earlier, WebRUSL is designed for the rapid processing and transmission of both data and results. The program helps companies set reorder points and reorder quantities for new spares and replenishment spares. Data requirements for each vary somewhat, as shown in Figure 10.3. To replenish items that are already in stock (and have at least one year of usage history), the necessary data parameters are usage, years of usage history, criticality of the part, lead time to replenish, average unit price, and set size (the number usually drawn at one time by maintenance personnel).

For new spares which typically lack usage history the number-in-service and the mean-time-between-failure (MTBF) are required. The number-in-service is usually available by checking bills of material for the new equipment. MTBF estimates tend to come from two sources: 1) vendor records (when you can get them to provide it), and 2) best es-

Requirements for rarely-used inventory:

Parameter	Description	
Usage	Total number of replacement parts issued (due to failure) during the period of historic data	Historic items only
Years data	The period (in years) covered by the usage data	
Criticality	The implication of getting caught short	
Lead time	The number of weeks to obtain, in routine replenishment, a replacement part that has been withdrawn from stock	
Average unit price	The average unit price of obtaining a part	
Set size	The number of units normally installed when item Is replaced	
Number in service	The number of installed applications that part will support	New spares only
Mean Time Between Failure (MBTF)	The length of time that the part is expected to perform between failures	

Figure 10.3 Replenishment and New Spares

timates from the plant maintenance and operating people.

Figure 10.4 shows all of the required fields in the file to be transmitted. In order to provide a valid estimate of the potential over- and understocks, client information on balance-on-hand, current MIN, MAX, and order quantity are also required.

The File Inventory Tree

We introduced the Inventory Tree in Chapter 1. Figure 10.5 shows an Inventory Tree for a client in the paper processing industry. Of the total 11,954 SKUs in the storeroom ($8,165,944 of inventory), 95.9 percent of the items and 92.3 percent of the inventory value were rarely used. To capture and analyze 78 percent of the total inventory value, only 2,048 RUKIs needed to be reviewed. These items were chosen to represent the majority of the inventory (typically 80-85 percent), while only requiring maintenance and operating personnel to set criticality

Field name	Data values
AUP	Required
LTIME	Required
STOCKNO	Required
DESC	Required
CRITCODE	Required
YRSDATA	Historic items only
USET	Historic items only
NOINSVC	New spares only
MTBF	New spares only
SETSOF	Optional
BACKIMP	Optional
BOH	Optional
CIP/MIN	Optional
CIP/MAX	Optional
CIP/OQ	Optional
PO/QTY	Optional

- All fields must be in the file uploaded to WebRUSL^{SM}

- Data values required will depend on if the query is for historic items or determining new spares

- Optional data values may be provided to assist you in comparing current settings and performance to WebRUSL^{SM} recommendations

- Add any fields that you want

- All fields can be in any order

Figure 10.4 WebRUSL Process – Required Fields

and set size for less than 20 percent of the items (Pareto's rule).

Although the RUKIs are the main focus of a project, conservative MIN/MAXs for the non-key items can easily be set, using default values for criticality and lead time, thereby avoiding the time and effort needed to determine more exacting values.

Avoiding Unnecessary Purchases

Two of the most effective ways for a client to cut near-term inventory cost is to: 1) avoid buying too many (excessive) new spares in the beginning, and 2) avoid buying more replacement spares than is necessary when parts reach the reorder point. (see Chapter 6 for a more detailed discussion of this subject.) Figure 10.6 shows a projection of

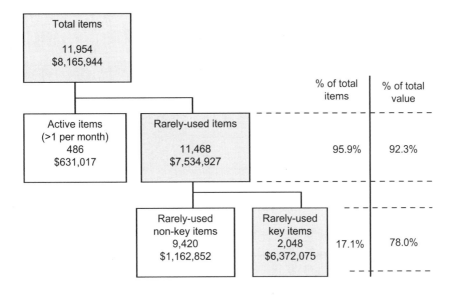

Figure 10.5 An Inventory Tree File Structure Analysis

how many RUKIs are probably in the procurement process now, and how many are expected to hit the reorder point over the next 24 months and after 24 months. These projections are based on current usage rates and current client MIN/MAX levels.

From its analysis, ISI has already determined that all of the 503 items indicated currently have reorder points that are likely set one unit, or more, too high. Deferring the re-purchase of the 503 items could possibly generate a cash flow saving of over $1.2 million. Some of the items may ultimately need to be procured after the two-year period, but many may never need to be procured because they are seldom, if ever, used.

Actual WebRUSL Results

Figures 10.7 and 10.8 show sample input and output files using the WebRUSL process. In this example, the client wanted to set MIN/MAXs for each of five items under two different lead times (26 and 13 weeks) and under two different criticality levels (high and low).

Benefits of re-evaluating replenishments – Some rarely used items have a high probability the MINs are too high and issuing a purchase order will only perpetuate or compound an overstock position

When replenishment is expected to occur	Number of key items	Percent of total key items(%)	Cost of unnecessary replenishment
Now	49	2.4	$239,783
Soon	158	7.7	$516,646
Later	296	14.5	$521,537
Totals	503	24.6	$1,277,966

MIN too high and up for reorder NOW

MIN too high and likely up for reorder In next 24 months

MIN too high and likely up for reorder after 24 months

Figure 10.6 Avoiding Unnecessary Purchases

Stock No.	AUP	Unit of issue	Lead time (Weeks)	Description	Criticality code	Set size	Years of data
100167	50.16	PR	26.0	Boot, Safety, Type	H	38	4.00
100167	50.16	PR	13.0	Boot, Safety, Type	L	38	4.00
102722	70.12	EA	26.0	Lighting fixture	H	19	4.00
102722	70.12	EA	13.0	Lighting fixture	L	19	4.00
108350	79.83	EA	26.0	Harness, Wiring	H	20	4.00
108350	79.83	EA	13.0	Harness, Wiring	L	20	4.00
110300	1391.26	EA	26.0	Filter Element	H	6	4.00
110300	1391.26	EA	13.0	Filter Element	L	6	4.00
110317	53.62	EA	26.0	Fuel Element	H	28	4.00
110317	53.62	EA	13.0	Fuel Element	L	28	4.00

Figure 10.7 A Sample Input File For WebRUSL

A total of four years of usage data were available.

Figure 10.8 shows the WebRUSL output file with the RUSL-recommended MIN/MAX/EOQ for each item. Normally, the MAX is set by adding the EOQ amount to the MIN. In this case, the calculation is different because the client reorders when the balance-on-hand for the spare reached one unit below the material system MIN. The analysis for part number 110300 indicates that the MIN/MAX/EOQ must increase from 2/2/1 to 5/5/1 as the lead time extends from 13 to 26 weeks and the criticality is increased from low to high (circled values).

10.2.2 BEST PRACTICE FOR "ACTIVE" ITEMS

The DAI Process Overview

Figure 10.9 outlines the overall process used by Decision Associates, Inc., (DAI) to set stocking levels for their client's active items. DAI uses its decision support software, called Stock Management System (SMS). Starting with the client input data, a forecast of future demand for each SKU is made by analyzing the data using 15 different forecasting models shown in Figure 10.10. The process for applying the forecast models is also discussed. Following this step, a safety stock level is determined to compensate for forecast error; the EOQ is then calculated for each item. Next, each SKU is sorted into one of seven ordering codes to indicate stock condition during the next lead time (Figure 10.11).

SMS can be linked to the client's computer to transmit and receive results automatically. Decision Associates algorithms can handle slow-moving items to some degree.

Applying The Forecasting Models

Because of the larger usage patterns of active items, it is possible to forecast fairly accurately future demand. Typically, the SMS forecasting process uses two years of recent usage history and applies the forecasting models as follows: 1) the usage data from the second oldest year is processed by each forecast model to generate a forecast for the

The RUSL MIN/MAX/EOQ varies based on criticality and lead time

Stock No.	AUP	Unit of issue	Lead time (Weeks)	Description	Criticality code	Set size	Years of data	RUSL Values		
								MIN	MAX	EOQ
100167	50.16	PR	26.0	Boot, Safety, Type	H	38	4.00	13	21	9
100167	50.16	PR	13.0	Boot, Safety, Type	L	38	4.00	6	14	9
102722	70.12	EA	26.0	Lighting fixture	H	19	4.00	9	14	6
102722	70.12	EA	13.0	Lighting fixture	L	19	4.00	4	9	6
108350	79.83	EA	26.0	Harness, Wiring	H	20	4.00	9	13	5
108350	79.83	EA	13.0	Harness, Wiring	L	20	4.00	4	8	5
110300	1391.26	EA	26.0	Filter Element	H	6	4.00	5	5	1
110300	1391.26	EA	13.0	Filter Element	L	6	4.00	2	2	1
110317	53.62	EA	26.0	Fuel Element	H	28	4.00	11	18	8
110317	53.62	EA	13.0	Fuel Element	L	28	4.00	5	12	8

Figure 10.8 A Sample Output File For WebRUSL

Decision Associates, Inc. (DAI), Chicago, IL USA

Starts process by performing a forecast of future demand
Uses 15 different forecasting models
Sets safety stock to compensate for forecast error
Calculates economic order quantity (EOQ)
Sets replenishment plan for next 52 weeks
Sets one of seven ordering codes to indicate stock condition
   during the item's lead time
Ordering codes range from extreme shortage to extreme excess
Separate screens show:
- forecast detail by stock number
- replenishment plan by quarter
- financial performance summaries

SMS can be linked to load results into material system
   automatically
No-cost trial of SMS for up to thirty days is available
DAI's algorithms can handle slow-moving items to some degree

Figure 10.9 A Best Practice for Active Items

most recent year of usage, 2) the forecast error for each model is determined, with the model having the lowest forecast error being assigned to each spare part, and 3) the assigned model is then used with both years of usage data to generate a forecast for the future 12 months. All forecasts are screened for bias and volatility (e.g., peaks).

Summarizing the Forecast

Figure 10.12 shows a printout of a typical forecast summary from SMS. Demand for the previous two years is shown along with the next year's raw forecast and the final forecast with any adjustments for bias or volatility (in this example, there were no adjustments). Also shown are the forecast for the next lead time (Forelead = 1578), the forecast error (11%), and the forecast model number selected (51=linear regression). In this example, a 95-percent service level was selected by the client for the roller bearing. From the total demand numbers at the bottom of the figure, it can be seen that the forecast for the roller bearing is falling each year.

SMS uses several forecasting models:
Moving average (3 Month/ 6 Month/ 12 Month)
Trend average (3 Month/ 6 Month/ 12 Month)
Seasonal (No Trend: This Year = Last Year)
(With Trend: 3 Month/ 6 Month/ 12 Month)
Single exponential smoothing (Alpha = 0.08/ 0.25/ 0.50)
Linear regression
Custom exponential smoothing (user sets Alpha)

The Process:
Each model produces a forecast for the most recent year, based on patterns from the prior year
Forecast error Is determined by comparing actual usage to predicted usage for the most recent year
All forecasts are screened and bias and volatility are eliminated
Remaining models are reviewed to select the one with lowest forecast error
The chosen model is used with the most recent year of usage to forecast next year

Figure 10.10 Forecasting Active Items With SMS

SMS uses these seven ordering codes to determine stock condition during the next lead time:

-3 = Existing outage

-2 = Projected outage

-1 = Normal reorder

0 = Balanced (No action required)

+1 = Overstocked

+2 = Overstocked with orders pending

+3 = One year excess

Figure 10.11 SMS' Seven Order Codes

Item 162234   Roller bearing   Location: Detroit

Forelead 1578

Error % 11

Model chosen 51

Model family 0

Sensitivity % 0

Alpha factor 0.00

Service level % 95

Inventory class

Lead time (Wks) 4

This forecast for next year had no adjustment

	Demand 2 years ago	Demand 1 year ago	Month	SMS raw forecast for next year	SMS adjusted forecast
	1760	1933	Mar	1707	1707
	4369	2321	Apr	1707	1707
	3348	1662	May	1664	1664
	2238	1798	Jun	1621	1621
	3458	1875	Jul	1579	1579
	2632	1948	Aug	1536	1536
	4156	1982	Sep	1493	1493
	1944	1982	Oct	1451	1451
	2892	1958	Nov	1408	1408
	2074	820	Dec	1366	1366
	2486	1272	Jan	1323	1323
	2072	2303	Feb	1280	1280
	Total 33,429	Total 21,854		Total 18,135	Total 18,135

Figure 10.12 A Sample Of A SMS Forecast For Active Items

Item 7648 Connector   Location: Detroit

	Date 3/24/04
Forelead	1578
On hand	0
Backorder	1,240
Price	0.87
Cost	0.17
Carry cost	0.190
Order cost	20.00
Lead time (wks)	4
Service level	95
Inv class	
Plan code ID	
# in use	
EOQ (actual)	4883
Safety stock	18

Timing	Forecast	Storeroom position start of week	Planned Additions	Storeroom position end of week
3/24	385	-1240	6508	4883
3/31	396	4883	0	4487
4/7	399	4487	0	4088
4/14	398	4088	0	3690
4/21	398	3690	0	3292
4/28	386	3292	0	2806
5/5	375	2806	0	2531
5/12	376	2531	4883	2155
5/19	376	2155	0	1779
5/26	376	1779	0	1403
6/2	378	1403	0	1025
6/9	379	5908	0	5529
Qtr 2	4592	5529	4883	-
Qtr 3	4232	5854	4883	-
Qtr 4	4050	5854	4883	-

Figure 10.13 A Sample Of A SMS Position Summary

The Position Summary

Figure 10.13 shows a position summary for a connector (Item 7648). Summary data including price, carrying cost factor, lead time, service level, EOQ, and safety stock level are shown on the left side. On-hand quantity (zero), the status of any backorders (1240 units in this example), and the projected usage during the next lead time (1578 units) are also shown. Weekly status is shown for the next quarter for the forecast, the storeroom position at the beginning of each week, planned additions, and the storeroom position at the end of each week; after that, quarterly values are shown. From the table it can be seen that the connector is in a backorder position the week of March 24, but recovers to a positive 4,883 units (an EOQ amount) after receipt of 6508 units and covering the week's forecast (385) and the backorder (1240).

The Overall Summary

Figure 10.14 shows three summaries: 1) the forecast performance summary, with the total units used in the most recent two years and the forecast for next year, 2) the financial status summary, which

Forecast performance summary: Total units

	Total units
Year 1	34,277
Year 2	32,714
Forecast	31,145

Financial status summary:

Current on-hand	Target investment	Dollar change	Carry cost	Target savings
$ 6,209,415	$ 4,614,462	$ 1,594,953	20%	$ 318,990

Order code distribution:

Back order	Out-of-stock	Normal	Balanced	Overstocked	Over w/orders	Surplus
-3	-2	-1	0	+1	+2	+3
7 %	17 %	1 %	55 %	7 %	1 %	11 %

Figure 10.14 Samples Of SMS Performance Summaries

shows the projected savings during the forecast period, and 3) the order code distribution. Decreasing the stocking level from the current on-hand value ($6,209,415) to the target investment of $4,614,462 will result in a $1,594,953 decrease, a savings of $318,990 using the 20 percent carry cost factor.

The order code distribution shows that 63 percent of the items in the analysis are in the +/-1 range indicating normal processing. Seventeen percent of the items are out-of-stock, with 7 percent on backorder. At the other end of the spectrum, 11 percent of the items show at least a one-year excess.

10.2.3 THE BEST OF THE BEST IMPLEMENTORS

This section will explain in detail why the Kansas City Board of Public Utilities (BPU) was selected as the best implementer of the RUSL process for slow-moving inventory.

Project Overview

For the last eight years, Inventory Solutions, Inc. (ISI) has helped BPU to optimize the stocking levels of spare parts for their power and water operations. Figure 10.15 summarizes the overall project for the 13 storerooms and 23,000 SKUs. Of the total inventory of $5,385,000, only 1,844 RUKIs worth $4,416,000 were selected for detail review.

The program was divided into three phases. During phase 1, criticality levels (high, medium, or low) were set by production personnel for the RUKIs. Once criticalities were set, data on parts usage, replenishment lead time, and other parameters affecting stock levels were used to calculate the optimum reorder point and reorder quantity. After approval by plant operating personnel, adjustments were made to out-of-balance items based on recommendations from the consultants. Modified ROPs and ROQs were then entered into the BPU computerized material control system. For simplicity, changes were made to the stocking levels for the non-key items without the need to set exact criticalities.

```
┌─────────────────────────────────────────────────────────┐
│            Kansas City, KS, Board Of Public Utilities (BPU)  │
│              A case study on how to implement effectively    │
│     Project overview:                                        │
│         13 storerooms with over $100,000 each of inventory   │
│         $5,385,000 total inventory                           │
│         23,000 SKUs                                          │
│         1,844 rarely-used key items (RUKIs) singled out for review │
│         $4,416,000 RUKI inventory                            │
│         Project divided into three phases:                   │
│                                                              │
│         Phase 1: Set reorder points for 1,844 RUKIs          │
│         Phase 2: Identify additional inventory reduction for │
│                  900 RUKIs by getting shorter supplier lead times │
│         Phase 3: Identify 7,700 non-key rarely-used items for │
│                  inventory reduction                         │
└─────────────────────────────────────────────────────────┘
```

Figure 10.15 The "Best" of the "Best"

Stocking of new spares was also included in the program. For these items, an estimated failure rate (along with criticality and number-in-service) was obtained from the supplier or plant maintenance personnel. BPU internal purchasing paperwork was modified to add new spares data entry.

Phase 2 identified additional inventory reduction by focusing on improving supplier replenishment lead times. Each RUKI was tested to determine if it was a candidate to submit to the vendor for verification of a shorter lead time to replenish (see Figure10.16). Only the RUKIs (about 900) meeting all of the following four criteria were submitted to the vendors:

- items where the current replenishment level was expected to hit in the next two years.
- items where at least $500 in inventory could be saved per item.
- items where at least one unit of inventory could be saved.
- items where lead-time reduction of more than fifty percent would not be required.

Criteria for items submitted to vendors:

- Items where the current replenishment level was
  expected to hit in the next two years

- Items where at least $500 of inventory could be saved

- Items where at least one unit of inventory could be saved

- Items where lead time reduction of more than fifty
  percent would not be required

Only 900 of the 1,844 RUKIs met all of the above criteria.
This phase of the project required two mailings, but got
a 95% response rate. Vendors failing to respond were
removed from future purchasing consideration.

Figure 10.16 Phase 2: Reducing Vendor Lead Time

A specific shorter lead time for each item was requested from the vendor. The new shorter lead time was calculated to cut the required inventory investment by the most dollars quickly and without sacrificing availability. This phase of the program required two separate mailing to vendors, which resulted in a 95 percent response rate. Vendors failing to respond were removed from future purchasing consideration.

Phase 3 was a relatively simple program that set MIN/MAXs for 7,700 non-key items using default values for criticality (high) and lead time (26 weeks). During this phase, a review by maintenance personnel of the items for set size was made.

Program Results

Throughout the program, BPU and ISI agreed to use the following three main measurements to track how closely the BPU inventory plan compared to the optimum inventory level:

- BPU total actual and planned inventory reduction.
- Absolute Variance Ratio (AVR).
- +/-1 percentage.

Planned and actual inventory reduction:

1,844 Key items

Phase 1	Actual starting inventory	$ 4,416,475
	Recommended ideal Inventory	$ 1,172,000
	BPU inventory plan after Phase 1	$ 1,222,000
Phase 2	Recommended ideal inventory level after Phase 2	$ 755,670
	BPU inventory plan after Phase 2	$ 932,780
Total	Actual and planned inventory reduction after Phases 1 and 2	$ 3,483,695

Figure 10.17 Results Of First Two Years Of BPU Project

The reader is encouraged to review Chapter 8 for a more detailed discussion of these metrics.

Once agreed-upon, ROP/ROQs were set for each item, the level of excess inventory was determined and the rate of work-off from normal usage was determined by year. Also, a significant number of items (about 389) were completely removed from inventory after being determined to be grossly overstocked or obsolete. Figure 10.17 shows the results from the first two years of the program.

Results during the first two years of the program were spectacular. Planned or actual inventory reductions programmed into the material system (ROP/ROQs adjusted) were $3,483,695. This reduction included benefits from contacting about 200 suppliers during phase 2; these suppliers committed to shorter delivery schedules for over seventy percent of the items requested. The shorter delivery times alone decreased the need to carry $416,330 of spares inventory.

A perfectly balanced inventory plan (no planned overstocks or planned understocks) would have an AVR of 0.00. BPU's AVR after phase 2 was 0.03 the lowest AVR ever achieved by any ISI client worldwide ever (see Figure 10.18). Equally impressive is the BPU +/-1 percentage of 99.5 percent. Only 0.5 percent of the RUKIs did not have

1,844 Key Items

Absolute Variance Rratio (AVR)	0.03
Percent of reorder points within ± 1 unit of recommendation	99.5%

The AVR determines the dollar amount by which stock items deviate from the ideal inventory level, both high and low

Figure 10.18 Key Measures Of BPU Project

their MIN within +/-1 unit of optimum for one reason or another.

Getting the Benefits Can Take Time

Slow-moving inventory takes time to adjust. It also takes time to get the momentum of the program going, criticalities set, lead times confirmed, and perhaps most important–getting the recommendations for adjustments to MIN/MAXs reviewed and accepted by operations and maintenance personnel. In theory, if a client is willing to accept a consultant's recommendations carte blanche and make the recommended adjustments in the material system, a perfectly balanced inventory plan can be achieved immediately. But that will probably never happen.

Look again at Figure 10.17. After phase 1, the recommended ideal inventory level was $1,172,000 for the 1,844 RUKIs. BPU came within $50,000 of accepting that level. After phase 2, the ideal was $755,670 and BPU's plan was $932,780, a $177,000 difference, reflecting operations unwillingness to accept many of the vendors' promises of shorter delivery.

Figure 10.19 shows actual and planned inventory over a ten-quarter period after initiating the program. Once the plan was set by quarter 3, it changed very little over the rest of the period. Figure 10.20 is even more interesting. It wasn't until quarter 6 that the best results were achieved, and these were not sustainable for very long. The delay in getting to the best results has already been mentioned mainly as a timing and momentum issue.

But why was the low AVR not sustained? There are several reasons: 1) recommended MIN/MAX levels changed somewhat each quarter after additional usage data was included in the analysis, 2) some lead times also changed, affecting the recommended stocking levels, and 3) order quantities did vary as prices fluctuated, causing the MAX stock level and average inventory values to change. Most of these factors came into play in the four quarters after quarter 6, but it was the judgment of the BPU management not to burden operations with requests to re-review the adjustments at that time. Cleaning up necessary adjustments were planned and are underway in 2005.

Why Was the BPU Program So Successful?

Figure 10.21 lists the four main reasons the BPU program was the

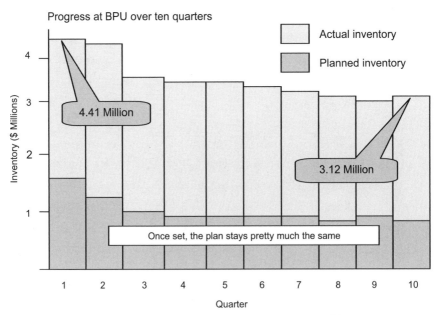

Figure 10.19 Changes In Actual And Planned Inventory

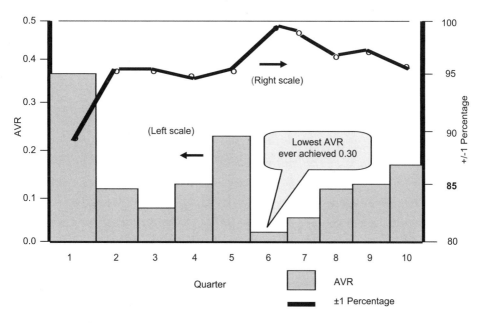

Figure 10.20 Progress At BPU Over Ten Quarters

The BPU project produced the best AVR and +/-1 percentage ever
for the following reasons:

Top management sponsored the project and reviewed the
results periodically

Plant maintenance and operating personnel participated in the
project data collection phase, then accepted nearly 90%
of the recommended adjustments

The software tools provided BPU were built into the procurement
process. All new spares and routine replenishment orders were
carefully reviewed before procurement

The management demanded acceptance of the program

Figure 10.21 Why the BPU Project was Successful

best implementation of all ISI clients. The first and last reasons listed are clearly the most important. Without the commitment and involvement of management, nothing of consequence would have happened. We will discuss this more in the following section under lessons learned.

10.3 IMPORTANT FINDINGS FROM YEARS OF CONSULTING

Figure 10.22 summarizes some important finding by the authors after over twenty years of consulting on the stocking of slow-moving inventory. Several of the findings deserve further comment.

The fact that 80-90 percent of production storeroom inventory is rarely used is not surprising considering that most of this inventory is

---

- 80-90% of all inventory in a typical MRO storeroom is rarely used

- Up to 40-45% of rarely-used inventory is tied up in overstocks

- About one-half of overstocked inventory will work off in 4 years

- About 40% of overstocked inventory is a result of first-time buys

- About 70% of initial buys are high, 10% are too low, and the rest are about right

- Spare parts inventory availability can be increased while overstocks are cut

- Lower-cost items are up to 10 times more likely to be overstocked

- Items with longer lead times are more likely to be understocked

- Vendor partnering can decrease required stocking by 20-30%

- Paretoís 80/20 principal works extremely will with inventory

- Shorter vendor lead times can cut required stocks by 20-30%

- Sharing spares among plants can cut required stocks by 20-35%

Figure 10.22 Important Findings For Slow-moving Inventory

for safety stock. What is somewhat surprising is that the 80-90 percent figure does not vary much, regardless of the industry studied. Refineries seem to have just as much rarely-used inventory as nuclear stations; steel mills just as much as paper mills.

The amount of overstocking found in plants in various industries does vary considerably. Utilities tend to carry somewhere between 35 and 45 percent overstocks, not surprisingly because they can usually include it in their rate base and get paid by the customer to carry it. Industrial plants, on the other hand, tend to control excess inventory levels to the 10-25 percent range, driven by the tough competition in their respective industries. The highest excess found (50-60 percent) was in the transportation industry in the United States and United Kingdom; the reason was probably related to government-operated monopolies.

Hundreds of analyses have consistently shown that about one-half of the slow-moving excess inventory will work off over a four- or five-year period if reorder points are adjusted and replenishment delayed. Remember that the yearly work-off forecasts are a moving target that can change as inventory parameters like demand and lead time change over time.

First-time buys of spares have been found to contribute to about 40 percent of the excess found in most storerooms, all the more reason to use a decision support tool of some kind to help set those initial buys. Relying on the vendor's recommendations is the major factor in 70 percent of the cases of initial over-buying. Remember that your vendor's spare part sales tend to carry higher margins than the original equipment.

Because of the cost bias observed in every storeroom, where the least expensive items tend to be carried at 8-10 times required levels, it is possible to improve inventory availability overall while cutting overstocks. Working off excess inventory can more than offset the higher inventory impact of setting the availability of highly-critical items at 99.9 percent (minimum).

Partnering with vendors and others can have significant impact on the amount of inventory carried by a storeroom, as discussed in Chapter 9. Inventory savings of 20-30 percent are achievable by keeping in touch with changes in lead times and other disruptions to the supply chain. Likewise, sharing stocks with other plants within and outside the company can cut required stocking levels by 20-30 percent easily.

10.4 IMPORTANT FINDING ABOUT IMPLEMENTATION

Earlier in this book, we discussed how Top 10 reports can help identify wide fluctuations in stocking values. We now introduce our top three reasons why many inventory programs are unsuccessful (see Figure 10.23).

Number 1: Maintenance Support

Unless there is strong support from maintenance personnel, projects to improve inventory are doomed to fail. In case study 8-3 (Chapter 8), we described how a top manager assigned inventory improvement as a shared goal between the plant managers and plant maintenance superintendents. We seldom attended a review meeting at those plants when both the plant manager and the maintenance superintendents were not present.

---

1. Without maintenance support, projects to improve inventory are doomed to fail

2. Best results occur when top management takes an active interest in monitoring results

3. Most inventory initiatives will be short lived unless internal procedures are modified to incorporate better inventory stocking procurement controls:

   • Require justification before purchasing
   • Verify the latest lead time before placing orders
   • Analyze new spares before procurement
   • Require suppliers to give information about Mean-Time-Before-Failure (MTBF) before buying original equipment

---

Figure 10.23 Implementing Inventory Programs

Unfortunately, we have had too many other occasions where projects were conceived at the corporate level and were either boycotted or ignored by the plant operations people as just one more intrusion from corporate. In most plants, maintenance either controls the stocking levels of spares or has veto power over any changes. Without their involvement in setting criticality and set size, and their input on accepting recommended changes, the project will not have lasting success.

Number 2: Top Management Involvement

The best results from an inventory initiative happen when top management takes an active interest in the program. It doesn't take much to make an impression on the staff – attending an occasional review session, or asking for a briefing from the consultants at a staff meeting. By showing a willingness to participate occasionally, top management is signaling to the staff that the program is important to them and, therefore, it should be important to the staff.

Number 3: Modify Internal Procedures

Most drives, projects, and campaigns have a finite life span. Many will die as soon as top management focuses on something else. Others may last for a few years until a new management appears on the scene, and then disappear without fanfare. But that need not happen if operating procedures are changed to import the best features of the inventory initiative into the routine procedures for managing the supply chain.

At BPU, for example, a new spare can not be procured until it is processed through the decision support tool and a recommended purchase amount is determined. It is unlikely that that process will change just because the General Manager moves on or the storeroom superintendent retires. The same is true for other simple rules such as requiring a lead time update on any spare over a certain dollar amount if the current lead time is more than two- or three years old. Like the old saying, "get it in writing," the rule for an inventory initiative becomes, "get it in the procedures."

- Senior management is seldom concerned about spare parts inventory until a crisis occurs that causes lost production

- Maintenance personnel usually express two major concerns:

    1. Your going to take away my inventory
    2. Your going to load me down with more work

- Stores personnel are especially sensitive to having outside consultants evaluate their storeroom inventory. Their major concern is they will be accused by management of not doing a good job if large amounts of excess inventory are found

- Managers seldom lose their job after making inventory improvements, but many lose them when they do *not* make improvements

- Most plants do not adhering to their own inventory plan

Figure 10.24 Challenges To Inventory Projects

10.5   IMPORTANT FINDINGS ABOUT MANAGING INVENTORY PROJECTS

10.5.1 Top Management Involvement

Let's face it. Inventory improvement does not usually head top management's list of priorities. Much of the time it's not even on the radar screen unless a spare part shortage has caused a production loss (see Figure 10.24). One reason operating personnel play it safe by grossly overstocking spares is to keep the heat off from upper management. Yet a politician once remarked about the federal budget, "save a few billion here, and a few billion there and, before you know it, you're talking real money." We may not be talking billions in investment savings, but saving a few thousand here or there can quickly add up to big bucks!

Over the years, we have been amazed by the almost total lack of interest in inventory projects by chief financial officers. During the hundreds of projects we have implemented across many industries, we can think of only one occasion when a CFO actually sponsored an inventory initiative. Some have even referred to the benefits of inventory initiatives as being in the rounding error of the numbers. Maybe so, but their plants are still carrying double-digit millions of dollars of unnecessary inventory that could be better employed elsewhere. If the shareholders only knew....

10.5.2 Maintenance Participation]

It's not surprising that maintenance personnel are skeptical when some new initiative is imposed on them, especially by the corporate staff. Having plenty of inventory in the storeroom is like a security blanket to them. They seldom see anything good coming from an inventory project because they equate most programs to taking away my inventory. One way we have found to offset their fear of not having enough inventory is to employ the concept of the incremental maximum as shown in Chapter 3 (Figure 3.22). Once maintenance workers see that adding additional inventory provides no additional improvement in spares availability, they usually agree that what they have is enough...at least most of the time.

10.5.3 Support from the Storeroom Management

Storeroom managers are especially sensitive to having outside consultants evaluate their storeroom inventory. Their major concern is that they will be accused by their management of not doing a good job if large amounts or excess inventory are found. This concern is unfortunate because seldom have most storeroom staffs had the benefit of any kind of decision support tool to help them manage the inventory better. Once they get over the fear factor and start getting involved in using the new tools available to them, things tend to improve quickly.

10.5.4 Who's Problem Is It?

Every company we have worked with has done a poor job of adhering to their own inventory plan. Remember: the inventory plan is set by the MIN/MAXs currently in the material system. Without excep-

Most companies do a poor job of adhering to their inventory plan

The following table is typical of most storeroom inventory situations

Stock number	Balance-on-hand (BOH)	Recommended MAX	Current storeroom MAX
0834695	17	1	2
0279956	83	3	5
1434554	7	1	2
7695857	4	1	1
0895564	26	2	2
1433428	13	4	7
6873647	9	1	1

Reasons why the BOH is over the current storeroom MAX:

We bought way too much when we built the plant. It never worked off!
The plant manager told us to buy it . . so we did!
We bought a lot for the overhaul, and didnít use it
We inherited most of It when our other plant closed
I have no Idea how the heck we got it!

Figure 10.25 Poor Performance Against Plan

tion, we have found in every storeroom hundreds of items having balances-on-hand that are in excess of the material system's maximum allowable limit. Some of the excuses are listed in Figure 10.25. Interestingly, you can never find anyone in the plant who will ever admit to being a party to the crime! "The plant manager told me to buy it" or "It was here when I was hired" are common excuses.

Once in a while, upper management will notice the problem and ask for a periodic listing of over-maximum items. Be careful here, however. Instead of fixing the problem, it is not uncommon for the material system maximums to be raised by someone trying to make the next month's list shorter. The material system's maximum stocking levels are like budget limits not to be exceeded without a justifiable reason.

CHAPTER 11

IMPLEMENTATION

11.1 WHAT THE READER SHOULD LEARN FROM THIS CHAPTER

- Key steps to having a successful project:
    - Develop a detailed schedule
    - Agree on roles and responsibilities
    - Review reports and take action to adjust stock levels
        Monitor results against goal
        Keep management informed
        Track implementation against plan
    - Suggested informational reports
    - Suggested action reports

11.2 INTRODUCTION

We have intentionally left this chapter on implementation until last for several important reasons: 1) it is useful to understand the nuances of setting MIN/MAX levels before tackling implementation, 2) the pros and cons of goal setting should be understood, and 3) knowing the best practices of others can be helpful when designing an implementation strategy.

In Chapter 10, we discussed some of the lessons we have learned during our years of consulting on inventory optimization. We stressed that most inventory initiatives are not going to be successful without the support of maintenance personnel. This is also true if top management does not show its support for the program. Furthermore, most implementations are short-lived unless internal procedures are changed to accommodate the company's new lifestyle introduced during the initiative.

For well over twenty years, we have assisted with inventory implementation programs at over 700 power stations and industrial plants around the world. On a scale of 1 to 10 (worst to best), we have rarely had an implementation that would rate a solid 10. The closest to it is the Kansas City, Kansas Board of Public Utilities described in Chapter 10 as the best of the best. Most implementations rated in the 5-to-7 range, and a few would barely even get a 2. Why? Mostly for the reasons just listed, especially the lack of top management support.

Upper management may sign-off on starting an initiative, but if they lose interest, so does everybody in the organization. A noted professor once asked his class, "If you could only read one journal or publication, which one would it be?" The students' answers ranged from the *Wall Street Journal* to *Fortune* magazine. "Wrong," said the professor, "The one publication you want to read is the one your boss reads." Not surprisingly, most people in an organization keep their attention focused on what the boss thinks is important.

In this chapter, we will discuss the actions we believe are important in order to implement your inventory initiative successfully. The process starts with getting an audience with the client to present a cost-benefit analysis that will convince them to fund a program. Subsequent steps include: 1) presenting the initial recommendations, 2) getting the client to review and approve some or all of the recommendations, 3) getting the approved changes into the material management system, 4) setting goals, and 5) monitoring the results against goals. Each of these areas will be discussed in detail in the following sections, including some sample reports for presenting results.

11.3   THE IMPLEMENTATION SEQUENCE

11.3.1 Getting The Data File

Getting a data file from a prospective client in order to perform a no-cost analysis should be easy. In reality, it is not. Companies are reluctant to hand over internal data to an outsider (consultant) unless there is serious pressure from upper management to fix something in the materials system. After that, the challenge can be one of timing —getting somebody in authority to say "yes." We have had many situations in our consulting career where we have worked on a client for more than five years before we even got close to selling a project. But we have learned to be persistent. Sometimes just as we are about to convince a senior manager to commit to a program, the manager moves to another job; we then have to start all over again with the replacement.

Determining the best way to reach the decision maker is itself not an easy task. We have tried them all: 1) direct mailings, 2) e-mailings, 3) cold calls, 4) advertising, 5) speaking at conferences, and 6) referrals. Probably the most effective of these is referrals, where a prospective client talks to an existing client who says good things about our work. Rarely, if ever, has a client agreed to pay anything up front to get an analysis of their inventory situation. Therefore, we do it for free, and even that is no guarantee of getting a data file.

Once the prospective client does agree to provide the data file, there are still problems to overcome. We send out a file structure letter that lists the required data fields needed. We also request current material systems parameters for the SKUs including usage history (2 to 3 years), current MIN/MAX values, balances-on-hand, part cost, unit of measure, description, and lead time. Seldom does a client have other key information in their material systems such as part criticality, set size, number of parts in service, and mean-time-before-failure (MTBF). Probably about 10-to-20 percent of the time, the initial data file received is not capable of being processed for lack of data or file formatting problems.

11.3.2 Selling The Project

If you think getting the data file was difficult, now comes the hard part: getting the contract. Our experience is that it takes at least two or more presentations to different levels of management to get a commitment to proceed. In more than twenty years of consulting, we have had only one or two contracts sold at the first presentation. The lowest-level manager in the management chain often chairs the first meeting. After getting a favorable recommendation at that level, several more meetings are required to get the final okay. Even if you succeed in getting a positive response all the way up the management chain, the project still has to wait for the inevitable budget approval. This step may take two or more years as your project competes with others for the limited discretionary funds available. But, with patience, the day may finally come when the project is approved. Now you have to deal with the lawyers.

For the most part, we have been successful in getting clients to accept our one-page licensing agreement that protects our proprietary software. In the vast majority of cases, clients have simplified the contracting process by merely issuing a purchase order that references our proposal, which contains the necessary pricing and schedule details. On occasion, contracting can be more extensive, with a lot of extra and, at times, excessive language thrown in by the client's attorneys. Our worst case was a client attorney who took our one-page license agreement and turned it into a ten-page tome.

A key to getting contracts for our service was providing the client with a no-cost minimum benefits analysis (MBA), as discussed earlier in Chapter 10. During the early years of our consulting business, we were able to get contracts from the major utilities without having to make too many concessions. Later, as our markets for consulting expanded to include the Fortune 500 industrial customers, we were forced to come up with more creative ways to sell projects. A good example is an approach we instituted about five years ago call PAYS (Pay-After-You-Save), where the client pays a modest upfront fee and the rest of the payout comes only after actual inventory reduction in the rarely-used key items is achieved. Almost all new projects now give the client the PAYS option. Nearly all like the approach.

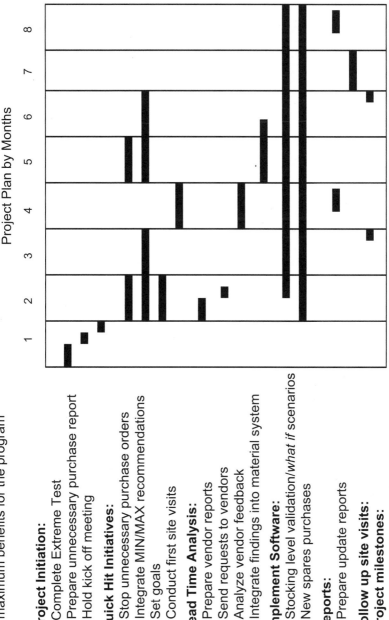

This project plan reflects the task and time line to implement process changes that achieve maximum benefits for the program

Project Plan by Months

Project Initiation:
  Complete Extreme Test
  Prepare unnecessary purchase report
  Hold kick off meeting

Quick Hit Initiatives:
  Stop unnecessary purchase orders
  Integrate MIN/MAX recommendations
  Set goals
  Conduct first site visits

Lead Time Analysis:
  Prepare vendor reports
  Send requests to vendors
  Analyze vendor feedback
  Integrate findings into material system

Implement Software:
  Stocking level validation/*what if* scenarios
  New spares purchases

Reports:
  Prepare update reports

Follow up site visits:
Project milestones:

Figure 11.1 Typical Project Schedule

11.3.3 The Project Schedule

Figure 11.1 shows a typical project schedule for the first eight months of a multi-plant project. The first steps in project initiation are to update the Extreme Test and hold a kick-off meeting to detail the project plan. High on the agenda are implementing the quick hit initiatives including: 1) stopping unnecessary purchases for items expected to be overstocked, 2) implementing MIN/MAX recommended changes for selected items, and 3) setting initial stock targets. In the schedule shown, a lead time analysis was included in the program to update purchasing lead times and to get vendor acceptance of shorter lead times for selected items. Figure 11.2 shows a more detailed explanation of the tasks.

Deliverable	Description
Extreme test	Complete analysis of MRO inventory, providing SKU detail reports for rarely-used key items (RUKIs), including: Recommended stocking levels (MIN/MAX/EOQ) Balance-on-hand greater than MAX (overstocks) Summary level cash flow – value of items that will work off over time
Avoid unnecessary purchases	SKU detail listing of rarely-used key items that have a high probability of MINs being too high and for which issuing a purchase order will only perpetuate or compound an overstock position
Lead time analysis	A process by which selected RUKI lead time's are updated through vendor participation, resulting in accurate current lead times which are then incorporated with actual criticalities to determine recommended MIN/MAX/EOQ values
Implement software	Log-on credentials, along with training on application use, provided to designated project team members; software available throughout the project for team members/sites to perform stock level validation, "what if" scenarios, and new spares stocking levels
Progress update reports	Comprehensive review of progress to date by site, at the SKU detail level, for all reports initially provided under the Extreme Test, as well as updated key indicators of performance for absolute variance ratio (AVR) and +/- 1 stocking point comparison
Site visits	Conduct indoctrination / training sessions for all project sites and divisions; task may be accomplished as either individual site visits or coordinated sessions with logical regional groups
Project milestones	Results are evaluated and compared against goals; schedules are re-evaluated and adjusted if necessary; management summaries are prepared; payment invoices are issued.

Figure 11.2 Task Descriptions

The roles and responsibilities of the various parties must be defined during the program. Figure 11.3 shows a typical assignment of responsibilities between the consultant and the client.

Although having a detailed schedule is important, adhering to the schedule can often be difficult due to emergencies and other considerations at the plant. When an unscheduled outage hits, just about everything else stops until production is restored. Experience has shown that schedules must be flexible, recognizing the need to accommodate unplanned events.

Task	Consultant	Client
Project initiation	• Complete Extreme Test analysis • Develop unnecessary purchase reports	• Provide data files • Identify project team members • Prepare and conduct kick off meeting
Quick hit initiatives	• Support development of project performance goals • Conduct initial site visits	• Implement unnecessary purchases and initial MIN/MAX recommendations • Coordinate initial site visits
Lead time analysis	• Prepare vendor reports • Provide communication template • Receive and analyze vendor feedback • Report findings	• Document communications • Compile vendor reply performance • Provide consultant with vendor reply information • Evaluate and implement findings
Implement software	• Provide technology as determined by project sponsor • Conduct training during site visits	• Designate project and site coordinators • Coordinate training with sites
Progress update reports	• Conduct and present analysis of data files	• Provide data files
Follow up site visits	• Conduct second site visit for review of progress and training	• Coordinate second site visit • Participate as required
Special analysis	• Conduct special analysis of rarely-used key items as identified	• Identify additional analysis opportunities related to project goals

Figure 11.3 Roles & Responsibilities

11.3.4 Presenting The Recommendations

When presenting results, we like to sort reports into Information Reports and Action Reports. Information reports are just that–they present useful information that will give the client key facts regarding the status of the project.

Action reports are the ones the client must review, accepting or rejecting the recommendations, and taking action to make the accepted changes in the material management system.

Information Reports – Figures 11.4 through 11.11 show a series of typical informational reports that can be delivered to clients after receiving every updated data file.

Figure 11.4: The Item Balance – This report shows the change in the actual and target inventory value for the rarely-used key items (RUKIs) from the last to the current data file. Also listed is the breakdown of any RUKIs that have disappeared from the last to the current file, items that have become active, and the value of obsolete items still in the material system. In the example shown, the actual inventory of the 1,030 current RUKIs increased from March 2003 to July 2004. The report does not identify the 28 items that are non-existing in the latest file. They can be available as a separate list.

Figure 11.5: Net Change In Actual Inventory – The net change in inventory value from both quantity impact and price impact are shown in Figure 11.5. In this example, inventory increased for some

Last Update: March, 2003   AJAX Manufacturing

RUKIs:	1,062	Obsolete items:	108
Inventory value :	$ 7,866,425	Value of obsoletes:	$ 938,373
Actual inventory :	$ 6,477,750		

	July 2004 Update			
	Mar 2003 RUKIs still RUKIs in Jul 2004	Mar 2003 items non-existing in Jul 2004	Mar 2003 RUKIs now active in Jul 2004	Mar 2003 items obsolete in Jul 2004
Number of items	1,030	28	4	107
Mar 03 target invy	$7,729,661	$121,832	$14,932	n/a
Mar 03 actual invy	$6,346,711	$100,750	$30,289	$938,373
Jul 04 target invy	$7,443,842	n/a	n/a	n/a
Jul 04 actual invy	$6,651,307	n/a	$56,814	$782,292
Change actual invy (decrease)	$ 304,596	($100,750)	$ 26,525	($156,081)

Figure 11.4 Typical Item Balance

```
                        AJAX Manufacturing

March 2003 actual inventory                          $6,346,711

        Quantity impact      (+)   1,414,031
        Quantity impact      (-)  (1,231,653)

Net quantity impact                  $182,378

Net actual before price impact                       $6,529,089

        Price impact         (+)       403,468
        Price impact         (-)      (282,250)

Net price impact                     $122,218

July 2004 actual inventory                           $6,651,307
```

Figure 11.5 Net Change In Actual Inventory

items by $1,414,031 and decreased for others by $1,231,653 (using constant prices from the March 2003 file). Likewise, price increases caused some items to increase in inventory value by $403,468, and for others decrease by $282,250. This report becomes important when monitoring inventory goals because it provides the basis for determining if the goal was obtained on a constant-dollar basis.

Figure 11.6: Typical Cost Bias Profile – We first discussed the concept of the cost bias in Chapter 2. Figure 11.6 shows a cost bias table for the 1,030 RUKIs under review at AJAX Manufacturing. The bias exhibited by AJAX is actually fairly mild, with an A/T (Actual/Target) of only 3.3 on the low cost end and 1.0 for items greater than $10,000. It is common to see biases of 8-10 and 0.5 for the low cost and high cost items, respectively. Because AJAX has set a current inventory plan (CIP) that is very low ($2,932,510) for the 1,030 RUKIs, the percent of the items in each cost range that is considered understocked (the last column) is very high. This AJAX plant has a serious potential understocking problem.

AJAX Manufacturing

Cost	Number of items	Target invy (T)	Actual invy (A)	CIP Invy	A/T	CIP/T	% under
< 100	26	11,023	36,374	7,063	3.3	0.6	80.8
100 < 250	38	41,904	106,222	11,932	2.5	0.3	78.9
250 < 500	92	158,053	187,144	65,931	1.2	0.4	68.5
500 < 1,000	147	363,359	458,572	129,605	1.3	0.4	67.3
1,000< 2,500	369	1,760,704	1,353,526	598,632	0.8	0.3	85.1
2,500< 5,000	236	2,228,694	1,758,169	820,913	0.8	0.4	87.7
5,000< 10,000	87	1,462,685	1,364,922	594,666	0.9	0.4	66.7
> 10,000	35	1,418,842	1,386,379	703,768	1.0	0.5	62.9
Totals	1,030	7,443,842	6,651,307	2,932,510	0.9	0.4	79.0

Figure 11.6 Typical "Cost Bias" Profile

Figure 11.7: Absolute Variance Ratio – AVR is one of the three key metrics we suggest clients use to monitor their performance against goal. Of the 1,030 items shown, 189 have exactly the reorder point and reorder quantity correctly set against the recommended values from the decision support software. Twenty-seven items have

AJAX Manufacturing

Target average inventory compared to CIP average inventory

Status	Number of items
CIP order point, order qty *same* as target order point, order qty	189
CIP planned avg inventory *greater* than target planned avg inventory	27
CIP planned avg inventory *less than* target planned avg inventory	814
Totals	1,030

Absolute Variance Ratio = 0.63

Figure 11.7 The Absolute Variance Ratio

MIN/MAXs set to overstock, and 814 are set to understock against recommended values. The resulting AVR of 0.63 is about normal when starting out in an inventory initiative. It will improve as changes are made in the MIN/MAXs.

Figure 11.8: Cash Flow Analysis – This report focuses on money. It shows how many of the 1,030 items are considered to be overstocked (241), by how much, and how fast the overstock value should work-off through usage each year for the next four years. Because these are slow-moving items, about one-half ($934,615) of the overstocked $1,809,584 will still be in the storeroom after four years.

AJAX Manufacturing

Rarely-used key items with actual balance-on-hand exceeding target maximum

	Current position			Cash flow based upon anticipated use ($)				
Store	Number of overstocks	Target Max invy ($)	Overstock value ($)	Year 1	Year 2	Year 3	Year 4	After 4
16	234	1,080,904	1,697,037	386,182	177,128	142,583	122,676	868,468
19	6	44,265	104,265	28,195	10,918	2,093	2,093	60,966
20	1	8,282	8,282	783	783	783	783	5,150
Total	241	1,133,450	1,809,584	415,160	188,828	145,459	125,522	934,615

Figure 11.8 The Cash Flow Analysis

Figure 11.9: +/-1 Percentage – This chart shows the results of our second metric for monitoring performance against goal. The +/-1 percentage measures the number of RUKIs that have their reorder point at, or within one unit +/-1 of the optimum (56.3 % in this case). As MINs are adjusted to get closer to the recommended MIN, the percentage will grow and approach 100 percent. A common goal is to get the number above 90 percent. This goal is not too difficult to achieve

AJAX Manufacturing
Comparison of CIP order point to suggested order point

Target - CIP	Number of items	% of total	Target invy ($)	Actual invy ($)	CIP invy ($)
-6	1	0.1	1,921	3,228	4,508
-5	1	0.1	2,271	2,753	4,509
-3	1	0.1	1,842	4,340	4,504
-1	18	1.7	66,236	307,292	134,740
0	205	19.9	938,081	2,163,784	961,452
+1	357	34.7	2,298,688	1,849,131	1,242,327
+2		14.6	1,227,685	898,107	390,002
+3			752,216	443,329	133,708
+4	56.3 percent within ±1 unit of optimum		626,641	240,401	78,520
+5			605,355	306,361	42,418
+6	37	3.6	410,392	180,115	(18,787)
+7	20	1.9	215,533	123,190	6,484
+8	13	1.3	141,494	69,620	3,828
+9	8	0.8	70,284	4,238	(10,287)
10 or More	4	0.4	85,202	55,419	(45,416)
Total	1,030	100.0	7,443,841	6,651,308	2,931,510

Figure 11.9 The +/-1 Percentage

because it can happen by simply changing the reorder point in the material system, it does not depend on excess inventory working off.

Figure 11.10: Balance-on-Hand Status. – There's an old saying in New York City that there is a story behind every light on a Broadway theater marquee. Similarly, we can find a story behind every item on this report where the current balance-on-hand exceeds not only the target MAX, but also the CIP MAX (Current Inventory Plan MAX). We have heard all of the reasons over the years: 1) "the plant manager told us to buy that much," (2) "we accepted the vendor's recommendation and now we've got too much," (3) "we bought it for an overhaul and never used it," and 4) the VP told the other plant to send it over here because he thought we could use it (we couldn't).

AJAX Manufacturing

(Abbreviated table)

Difference ($)	Balance-on-hand	Target MAX	CIP MAX	Part No.	Description	Avg unit price ($)
153,388	21	2	2	99876	Liner set, angle valve, 8X5	8,020.42
88,089	10	3	2	88786	Liner set, valve 12X10,SS	12,584.08
65,000	14	1	2	33547	Press duct without flange	5,000.00
54,986	3	1	2	99730	Pump, hydraulic, model 98	27,492.96

Number of records through page 1	= 36	Value through page 1	= 1,099,707
Total number of records	= 241	Total value	= 1,809,584
Percent of total records through page1	= 14.9%	Percent through page 1	= 60.8

Figure 11.10 Balance-on-Hand Status

Regardless of how the balances got to be so much higher than the system MAX, the fact that they are that high suggests a serious lack of discipline in the management of the supply-chain system at plants. Although at times it can be blamed on the "great quantity discount we got," that problem is seldom the case. This phenomenon of having too much inventory over the system MAX has been observed at virtually every storeroom we have ever analyzed, the only difference being the magnitude of the overstocking.

Figure 11.11: Actual Inventory Greater Than Target Maximum – If you thought the Balance-on-Hand over CIP MAX report showed a serious lack of discipline by AJAX, consider the report of actual inventory greater than target maximum. This report not only shows items that are over the target MAX, but looks especially at items that have gone over target since the project stated. Look at the first item (Part No. 22341). The on-hand quantity in March 2003 was 14 units, but by July 2004 it had increased to 80 units against a target MAX of 8 units.

There can be numerous explanations for this: 1) we conducted a cycle count and found all the extras, 2) they were drawn out for an over-

(Abbreviated table)   AJAX Manufacturing

Part No.	Description	Price ($)	Mar 03 on-hand	Mar 03 target MAX	Jul 04 on-hand	Difference ($)
22341	Sleeve, shaft	606.33	14	8	80	40,017
80355	Pump, hydraulic	27,492.96	2	2	3	27,492
78599	Roller, fairlead	220.00	33	18	52	4,180
93897	Deflector, liner	1,500.00	5	3	10	7,500

Number of records through page 1   = 33   Value through page 1   = 235,103
Total number of records   = 33   Total value   = 235,103

Figure 11.11 Actual Inventory Greater Than Target Maximum

haul, not used, and came back into the storeroom after the benchmark file was prepared, 3) we inherited them from another plant, 4) they were drawn out to fix a failure and the failed parts were repaired and returned to the storeroom, and 5) we bought more by mistake. Regardless of the reason, too much inventory was added to the storeroom and may never work-off. Remember, this increase occurred while the focus was on improving inventory optimization. Guess what happens when nobody is looking!

Action Reports

Action reports recommend specific actions that need to be taken to improve inventory performance. In all cases, they are item-specific. We suggest focusing on only the four illustrated in Figures 11.12 through 11.15.

Figure 11.12. Action Report For Overstocks. – This is a rank-order report showing the recommended adjustment to the MIN, MAX, and OQ for items where one or more of these settings are considered too high for the criticality assigned to the item. The column for the AVR Index shows the decrease in inventory value expected if the current values are adjusted to the target values. Notice Part No. 12221.

AJAX Manufacturing

Current MIN or order quantity needs adjusting

Part No.	Criticality	Current values				Target values			AVR index
		On hand	MIN	MAX	OQ	MIN	MAX	OQ	
24317	High	3	2	3	1	2	4	2	3,918
09588	High	12	12	24	12	6	11	5	2,318
12221	High	9 *	6	12	6	3	5 m	2	2,179
05888	High	9	3	4	1	3	5	2	2,009
86004	High	10	12	24	12	7	13	6	1,963

Number of records through page 1	= 22	Value through page 1 = 23,498
Total number of record	= 37	Total value = 24,250
Percent of total records through page 1	= 59.5%	Percent through page 1 = 96.9%

Figure 11.12 Action Report For Overstocks

The asterisk * that follows the on-hand quantity indicates that the part is likely to come up for routine replenishment within the next 12 months; the m behind the target MAX indicates that the item is issued from the storeroom as a set, in this case 2 units. As the items on this report are reviewed, the adjustments accepted, and the changes make in the material system, the AVR and the +/-1 ratio will both improve.

AJAX Manufacturing

High criticality items where current MIN needs to be adjusted upward

Part No.	Description	Price ($)	BOH	Target ROP	CIP ROP	CIP Availability	Increase from	Increase to	New Avail
00280	Lock	153.06	10	7	5	98.87	5	7	99.91
00553	Block	29.49	60	1	0	99.87	0	1	99.99
03289	Manifold	242.56	8	10	1	34.21	1	7	99.05
66364	Roller	220.00	52	6	4	99.29	4	6	99.97

Number of records through page 1   =   38
Total number of records   = 379
Percent of total records through page 1   = 10.0%

Figure 11.13 Action Report For Understocks (No Cost Adjustments)

Also, the item will disappear off the report the next time. Shorter lists are better.

Figure 11.13: Action Report For Understocks – There are three reports for understocks, one each for high criticality, medium criticality, and low criticality items. Because the minimum acceptable availability for a highly-critical item is 99.9 percent, Part No. 00280 does not meet that standard at the current reorder point of 5 units. To meet the 99.9 standard, the CIP ROP must be raised to 7 units. As it turns out, the balance-on-hand is currently at 10 units. Therefore, this adjustment can be made in the material system without purchasing additional inventory (hence the term No-Cost Adjustments). For every item shown in the report, the adjustment can be made without cost. If the balance-on-hand is not large enough to cover the entire recommended change, only a partial adjustment can be made.

Figure 11.14: Action Report For "Do Not Stock" Items – This report lists items that do not need to be stocked because they are low criticality, have a short lead time, and no usage. Not stocking them will still meet the 97.0 minimum available for low criticality. Usually these items are locally available from a distributor.

AJAX Manufacturing

Items no longer needs to be stocked at plant

Part No.	Description	Criticality	Price ($)	Lead time (Weeks)	Average annual usage	CIP Invy ($)
17373	Motor, Size A	Low	455.00	1	0.0	455
22044	Impeller, Pump	Low	270.83	1	0.0	270
00583	Valve, Kit	Low	466.00	1	0.0	466

Number of records through page 1 = 4
Total number of records = 4
Percent of total records through page 1 = 100.0%

Figure 11.14 Action Report For "Do Not Stock" Items

Figure 11.15: Plant Feedback Report – This report is used to summarize the actions taken by a plant on the latest set of recommendations from the consultant. It shows the number of RUKIs on each report, the number that have been accepted and had the MINMAX adjusted, and the number that have been reviewed and disapproved for adjustment. When the latter items are identified to the consultant they are removed for consideration in future reports. Notice that it is not necessary for the plant to predict the effect of the adjustments made on inventory value, AVR, or the +/-1 percentage because these metrics will be determined by the consultant after receiving the next data file.

Date submitted: Feb 5 2005   AJAX Manufacturing

Report to be reviewed	Number of RUKIs	Number of MIN/MAX adjustments made	Number of RUKIs reviewed but not adjusted (attach list)
Overstocks (MIN or OQ too high)	58	36	22
Understocks – High Criticality (MIN too low)	36	34	2
Understocks – Medium Criticality (MIN too low)	29	29	0
Understocks – Low Criticality (MIN too low)	1	1	0
Actual Invy Over Target MAX	35	26	9

Plants should complete this report within 45 days of receiving the latest consultants quarterly report; items reviewed but not adjusted will be removed from future reports.

Figure 11.15 Plant Quarterly Feedback Report

11.3.5 Reviewing The Recommendations

Over the years we have seen clients try various approaches for reviewing recommended changes. By far the best and most effective way has been to assign a team of maintenance and storeroom personnel to review the progress, preferably the same people who originally set the criticalities. One midwestern utility client has it down to a science. Each Tuesday (one hour before quitting time) the team meets to set criticalities for as many items as they can (usually about 50 to 60). The next day, the stores manager processes the items through the Internet software to determine the recommended MIN/MAXs. The team returns for one hour on Thursday to accept or reject the recommended values, and by Friday the accepted changes are entered into the material management system. Over a four-month period the team was able to completely set criticalities, review, and change all of the key items (about 800 SKUs). The key to this smooth process was setting up the twice-per-week schedule and adhering to it. With some clients, there has often been a long delay before the items reviewed are actually entered into the material system. This can be unfortunate, causing unnecessary purchases to be made until the old stock levels are corrected.

11.3.6 Monitoring Performance Against Goals

We discussed setting goals extensively in Chapter 8. It is important to monitor performance against goals in the implementation phase. Figure 11.16 shows a report for one of the AJAX Manufacturing plants. The change in values for all three of the primary goals are shown over a one-year period. Notice that the inventory change went down initially (line [2]–[3]) and then increased over the next two quarters due to a surge in approved purchases. During the same period, the AVR steadily decreased (lower is better) and the +/-1 percentage increased (higher is better), indicating that the overall inventory balance was steadily improving. Two other metrics are also reported: 1) the net change in prices, and 2) the amount of inventory added to the storeroom that was over the recommended MAX.

11.3.7 The Action Plan

It is always useful to have a method of tracking what everybody is suppose to do to keep the implement plan working on schedule. Figure

AJAX Aruba

Item	File Dates			
	Benchmark (Jul03)	Update (Nov03)	Update (Mar04)	Update(Jul04)
Months since start	-	5	9	14
Benchmark RUKIs	2,170	2,170	2,170	2,170
Benchmark RUKI value ($)	[1] 5,416,123	5,416,123	5,416,123	5,416,123
Update RUKIs	2,170	2,168	2,059	2,058
Update RUKI benchmark ($)	[2] 5,416,123	5,399,409	4,450,161	4,447,461
Update RUKI current ($)	[3] 5,416,123	5,311,367	4,451,237	4,651,307
Inventory change ($) (XXX) = decrease	[1]-[2] 0 [1]-[3] 0 [2]-[3] 0	(16,714) (104,756) (88,042)	(965,962) (964,886) 1,076	(968,662) (764,816) 203,846
Inventory value over target MAX since benchmark ($)	-	185,097	165,886	135,104
Net price changes since benchmark file ($) (XXX) = decrease	-	(44,997)	25,411	122,218
Absolute Variance Ratio (AVR)	-	0.68	0.54	0.42
% RUKIs within +/-1 unit Of target MIN	-	53.0	69.1	81.4

Figure 11.16 Quarterly Status Against Goal

PROJECT ACTION PLAN 2004/2005 *AJAX PAPER* *2rd QTR 2004*	
Client Contact: Lee Mason	Consultant: Bob Smith
Date completed	Consultant will . . .
4/15/04 4/18/04 4/26/04 4/30/04	1. Re-run excess report for last Qtr. 2. Recalculate AVR 3. Adjust RUKIs to remove dupes 4. Run MIN/MAX report 5. Run 3-table IRIX report 6. Prepare new recommendations 7. Run new spares as requested 8. Re-run new MBA report
Date completed	Client will . . .
4/12/04	1. Check status of 33 obsolete items 2. Eliminate duplicates 3. Issue planned PO adjustments 4. Send fresh files in July and Dec

Figure 11.17 Tracking The Implementation

11.17 shows a simple method for tracking the status against actions for both the client and the consultant. As actions are completed, they are noted in the "Date Completed" column.

11.3.8 A Few Final Comments

There's an old saying in the real estate business that the three main things in selling a piece of property are "location, location, and location." Similarly, there are three main factors necessary to having a successful inventory initiative; they are "implementation, implementation, and implementation." We hope that in this chapter we have given the reader a few good ideas on how to implement better. Equally important, we hope that throughout the book, we have provided some good insights on how to optimize your MRO inventory asset. As consultants we believe that's our job.

Centrifugal pump

INDEX